Mushroom Identification Logbook

A guided record book for the wild mushroom hunter

First paperback edition October 2019

Cover Photograph by Koldunov Photography

ISBN-10: 1698819277
ISBN-13: 978-1698819273

amazon.com/author/natureloversessentials

How to use this book

This logbook was created to help the mushroom forager carefully record key characteristics essential for mushroom identification. It is intended to be a companion to your favorite mushroom field guide(s).

Hand drawn illustrations and lists of possible characteristics will assist and remind the amateur mycologist of the essential features to record. Documenting detailed observations *before* reading mushroom descriptions will help you reduce preconceptions during the identification process.

Blank pages are provided at the end of the book as a space to sketch a particular interesting specimen, write extra notes, create spore prints (if space allows) or to draw a map of a particular abundant area. Once this book is completed you will have a record of your journey as a mushroom forager and possibly a reference to a bountiful collection area.

Mushroom Identification Log

Location

Site ______________________ **Date** ____________

◯ living tree ◯ leaf litter ◯ mulch ◯ other mushrooms
◯ dead tree or wood ◯ grass ◯ soil ◯ other ________

Type of tree(s) on or near ______________________

Forest type ______________________

Other (weather, lighting, humidity, etc.): ______________________

____

General

Size (cap, stem, & overall height): ______________________

Texture: tough, brittle, leathery, woody, soft, slimy, spongy, powdery, waxy, rubbery, watery, other: ____________

______________________ Color ____________

Spore Color ______________________

Cap Shape & Features

bell-shaped
conical
convex
cylindrical
depressed
flat
funnel-shaped
umbilicate
umbonate

Surface texture (warts, scales, slime, etc.) ____________

____

Cap margin: smooth, inrolled, sinuous/wavy, other:

____

Color changes ______________________

Gill Features

◯ gills ◯ false gills ◯ pores ◯ teeth

Spacing

crowded
close
distant

Gill Attachment

adnexed
(narrowly attached)

adnate
(broadly attached)

decurrent
(running down the stalk)

free
(gills don't meet the stem)

sinuate
(notched)

sketch

Stem Shape

equal

club-shaped

bulbous

with cup
(volva)

tapering
downward

tapering
upward

sketch

Other Features

Stem attachment ◯ central ◯ offset ◯ lateral

Stem texture: fleshy, fibrous, hollow, solid, fibrillose, scaly, velvety, smooth, other: ____________________

Veil ◯ none ◯ partial ◯ universal

Veil texture: thin, cobwebby, fibrillose, slimy, other: ________

____

Annulus (ring)

◯ none ◯ collarlike

◯ skirtlike ◯ sheathlike

Volva

◯ none ◯ present

Other (bruising, bleeding, smell, etc.) ____________________

____

Mushroom Identification Log

Location

Site ______________________ **Date** __________

◯ living tree ◯ leaf litter ◯ mulch ◯ other mushrooms

◯ dead tree or wood ◯ grass ◯ soil ◯ other ________

Type of tree(s) on or near ______________________

Forest type ______________________

Other (weather, lighting, humidity, etc.): ______________________

____

General

Size (cap, stem, & overall height): ______________________

Texture: tough, brittle, leathery, woody, soft, slimy, spongy, powdery, waxy, rubbery, watery, other: ______________________

______________________ Color __________

Spore Color ______________________

Cap Shape & Features

bell-shaped

conical

convex

cylindrical

depressed

flat

funnel-shaped

umbilicate

umbonate

Surface texture (warts, scales, slime, etc.) ______________________

____

Cap margin: smooth, inrolled, sinuous/wavy, other:

____

Color changes ______________________

Gill Features

◯ gills ◯ false gills ◯ pores ◯ teeth

Spacing

crowded close distant

Gill Attachment

adnexed
(narrowly attached)

adnate
(broadly attached)

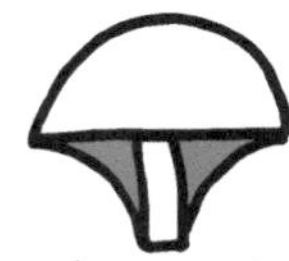

decurrent
(running down the stalk)

free
(gills don't meet the stem)

sinuate
(notched)

sketch

Stem Shape

equal

club-shaped

bulbous

with cup
(volva)

tapering
downward

tapering
upward

sketch

Other Features

Stem attachment ◯ central ◯ offset ◯ lateral

Stem texture: fleshy, fibrous, hollow, solid, fibrillose, scaly, velvety, smooth, other: ______________________________

Veil ◯ none ◯ partial ◯ universal

Veil texture: thin, cobwebby, fibrillose, slimy, other: ______

____

Annulus (ring)

◯ none ◯ collarlike
◯ skirtlike ◯ sheathlike

Volva

◯ none ◯ present

Other (bruising, bleeding, smell, etc.) ______________________

____

Mushroom Identification Log

Location

Site ______________________ **Date** __________

◯ living tree ◯ leaf litter ◯ mulch ◯ other mushrooms
◯ dead tree or wood ◯ grass ◯ soil ◯ other __________

Type of tree(s) on or near ______________________

Forest type ______________________

Other (weather, lighting, humidity, etc.): ______________________

____

General

Size (cap, stem, & overall height): ______________________

Texture: tough, brittle, leathery, woody, soft, slimy, spongy, powdery, waxy, rubbery, watery, other: __________

______________________ Color __________

Spore Color ______________________

Cap Shape & Features

bell-shaped
conical
convex
cylindrical
depressed
flat
funnel-shaped
umbilicate
umbonate

Surface texture (warts, scales, slime, etc.) __________

____

Cap margin: smooth, inrolled, sinuous/wavy, other:

____

Color changes ______________________

Gill Features

◯ gills ◯ false gills ◯ pores ◯ teeth

Spacing

crowded
close
distant

Gill Attachment

adnexed
(narrowly attached)

adnate
(broadly attached)

decurrent
(running down the stalk)

free
(gills don't meet the stem)

sinuate
(notched)

sketch

Stem Shape

equal

club-shaped

bulbous

with cup
(volva)

tapering
downward

tapering
upward

sketch

Other Features

Stem attachment ◯ central ◯ offset ◯ lateral

Stem texture: fleshy, fibrous, hollow, solid, fibrillose, scaly, velvety, smooth, other: ______________________

Veil ◯ none ◯ partial ◯ universal

Veil texture: thin, cobwebby, fibrillose, slimy, other: ________

____

Annulus (ring)

◯ none ◯ collarlike

◯ skirtlike ◯ sheathlike

Volva

◯ none ◯ present

Other (bruising, bleeding, smell, etc.) ______________________

____

Mushroom Identification Log

Location

Site ______________________ **Date** __________

◯ living tree ◯ leaf litter ◯ mulch ◯ other mushrooms
◯ dead tree or wood ◯ grass ◯ soil ◯ other ______

Type of tree(s) on or near ______________________

Forest type ______________________

Other (weather, lighting, humidity, etc.): ______________________

____

General

Size (cap, stem, & overall height): ______________________

Texture: tough, brittle, leathery, woody, soft, slimy, spongy, powdery, waxy, rubbery, watery, other: ______________________

______________________ Color ______________

Spore Color ______________________

Cap Shape & Features

bell-shaped

conical

convex

cylindrical

depressed

flat

funnel-shaped

umbilicate

umbonate

Surface texture (warts, scales, slime, etc.) ______________________

____

Cap margin: smooth, inrolled, sinuous/wavy, other:

____

Color changes ______________________

Gill Features

◯ gills ◯ false gills ◯ pores ◯ teeth

Spacing

crowded

close

distant

Gill Attachment

adnexed
(narrowly attached)

adnate
(broadly attached)

decurrent
(running down the stalk)

free
(gills don't meet
the stem)

sinuate
(notched)

sketch

Stem Shape

equal

club-shaped

bulbous

with cup
(volva)

tapering
downward

tapering
upward

sketch

Other Features

Stem attachment ◯ central ◯ offset ◯ lateral

Stem texture: fleshy, fibrous, hollow, solid, fibrillose, scaly, velvety, smooth, other: ____________________

Veil ◯ none ◯ partial ◯ universal

Veil texture: thin, cobwebby, fibrillose, slimy, other: ________
____

Annulus (ring)
◯ none ◯ collarlike
◯ skirtlike ◯ sheathlike

Volva
◯ none ◯ present

Other (bruising, bleeding, smell, etc.) ____________________
____

Mushroom Identification Log

Location

Site ______________________ **Date** __________

◯ living tree ◯ leaf litter ◯ mulch ◯ other mushrooms
◯ dead tree or wood ◯ grass ◯ soil ◯ other ________

Type of tree(s) on or near ______________________

Forest type ______________________

Other (weather, lighting, humidity, etc.): ______________________

____

General

Size (cap, stem, & overall height): ______________________

Texture: tough, brittle, leathery, woody, soft, slimy, spongy, powdery, waxy, rubbery, watery, other: ______________________

______________________ Color ______________________

Spore Color ______________________

Cap Shape & Features

bell-shaped

conical

convex

cylindrical

depressed

flat

funnel-shaped

umbilicate

umbonate

Surface texture (warts, scales, slime, etc.) ______________________

____

Cap margin: smooth, inrolled, sinuous/wavy, other:

____

Color changes ______________________

Gill Features

◯ gills ◯ false gills ◯ pores ◯ teeth

Spacing

crowded

close

distant

Gill Attachment

adnexed
(narrowly attached)

adnate
(broadly attached)

decurrent
(running down the stalk)

free
(gills don't meet the stem)

sinuate
(notched)

sketch

Stem Shape

equal

club-shaped

bulbous

with cup
(volva)

tapering
downward

tapering
upward

sketch

Other Features

Stem attachment ◯ central ◯ offset ◯ lateral

Stem texture: fleshy, fibrous, hollow, solid, fibrillose, scaly, velvety, smooth, other: ______

Veil ◯ none ◯ partial ◯ universal

Veil texture: thin, cobwebby, fibrillose, slimy, other: ______

______

Annulus (ring)

◯ none ◯ collarlike
◯ skirtlike ◯ sheathlike

Volva

◯ none ◯ present

Other (bruising, bleeding, smell, etc.) ______

______

Mushroom Identification Log

Location

Site ______________________ **Date** ____________

◯ living tree ◯ leaf litter ◯ mulch ◯ other mushrooms
◯ dead tree or wood ◯ grass ◯ soil ◯ other ________

Type of tree(s) on or near ______________________

Forest type ______________________

Other (weather, lighting, humidity, etc.): ______________________

____

General

Size (cap, stem, & overall height): ______________________

Texture: tough, brittle, leathery, woody, soft, slimy, spongy, powdery, waxy, rubbery, watery, other: ______________________

______________________ Color ____________

Spore Color ______________________

Cap Shape & Features

bell-shaped

conical

convex

cylindrical

depressed

flat

funnel-shaped

umbilicate

umbonate

Surface texture (warts, scales, slime, etc.) ______________________

____

Cap margin: smooth, inrolled, sinuous/wavy, other:

____

Color changes ______________________

Gill Features

◯ gills ◯ false gills ◯ pores ◯ teeth

Spacing

crowded

close

distant

Gill Attachment

adnexed
(narrowly attached)

adnate
(broadly attached)

decurrent
(running down the stalk)

free
(gills don't meet the stem)

sinuate
(notched)

sketch

Stem Shape

equal

club-shaped

bulbous

with cup
(volva)

tapering downward

tapering upward

sketch

Other Features

Stem attachment ◯ central ◯ offset ◯ lateral

Stem texture: fleshy, fibrous, hollow, solid, fibrillose, scaly, velvety, smooth, other: ____________

Veil ◯ none ◯ partial ◯ universal

Veil texture: thin, cobwebby, fibrillose, slimy, other: ____________

____

Annulus (ring)

◯ none ◯ collarlike

◯ skirtlike ◯ sheathlike

Volva

◯ none ◯ present

Other (bruising, bleeding, smell, etc.) ____________

____

Mushroom Identification Log

Location

Site ______________________ **Date** __________

◯ living tree ◯ leaf litter ◯ mulch ◯ other mushrooms
◯ dead tree or wood ◯ grass ◯ soil ◯ other __________

Type of tree(s) on or near __________

Forest type __________

Other (weather, lighting, humidity, etc.): __________

____

General

Size (cap, stem, & overall height): __________

Texture: tough, brittle, leathery, woody, soft, slimy, spongy, powdery, waxy, rubbery, watery, other: __________

__________ Color __________

Spore Color __________

Cap Shape & Features

bell-shaped
conical
convex
cylindrical
depressed
flat
funnel-shaped
umbilicate
umbonate

Surface texture (warts, scales, slime, etc.) __________

____

Cap margin: smooth, inrolled, sinuous/wavy, other:

____

Color changes __________

Gill Features

◯ gills ◯ false gills ◯ pores ◯ teeth

Spacing

crowded
close
distant

Gill Attachment

adnexed
(narrowly attached)

adnate
(broadly attached)

decurrent
(running down the stalk)

free
(gills don't meet the stem)

sinuate
(notched)

sketch

Stem Shape

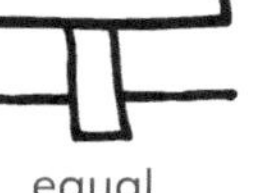

equal

club-shaped

bulbous

with cup
(volva)

tapering
downward

tapering
upward

sketch

Other Features

Stem attachment ◯ central ◯ offset ◯ lateral

Stem texture: fleshy, fibrous, hollow, solid, fibrillose, scaly, velvety, smooth, other: ______________________

Veil ◯ none ◯ partial ◯ universal

Veil texture: thin, cobwebby, fibrillose, slimy, other: ________

____

Annulus (ring)

◯ none ◯ collarlike

◯ skirtlike ◯ sheathlike

Volva

◯ none ◯ present

Other (bruising, bleeding, smell, etc.) ______________

____

Mushroom Identification Log

Location

Site ______________________ Date ____________

◯ living tree ◯ leaf litter ◯ mulch ◯ other mushrooms
◯ dead tree or wood ◯ grass ◯ soil ◯ other________

Type of tree(s) on or near ______________________

Forest type ______________________

Other (weather, lighting, humidity, etc.): ____________

____

General

Size (cap, stem, & overall height): ____________

Texture: tough, brittle, leathery, woody, soft, slimy, spongy, powdery, waxy, rubbery, watery, other: ____________

______________________ Color ____________

Spore Color ______________________

Cap Shape & Features

bell-shaped

conical

convex

cylindrical

depressed

flat

funnel-shaped

umbilicate

umbonate

Surface texture (warts, scales, slime, etc.) ____________

____

Cap margin: smooth, inrolled, sinuous/wavy, other:

____

Color changes ______________________

Gill Features

◯ gills ◯ false gills ◯ pores ◯ teeth

Spacing

crowded

close

distant

Gill Attachment

adnexed
(narrowly attached)

adnate
(broadly attached)

decurrent
(running down the stalk)

free
(gills don't meet the stem)

sinuate
(notched)

sketch

Stem Shape

equal

club-shaped

bulbous

with cup
(volva)

tapering
downward

tapering
upward

sketch

Other Features

Stem attachment ⬭ central ⬭ offset ⬭ lateral

Stem texture: fleshy, fibrous, hollow, solid, fibrillose, scaly, velvety, smooth, other: ______

Veil ⬭ none ⬭ partial ⬭ universal

Veil texture: thin, cobwebby, fibrillose, slimy, other: ______

______

Annulus (ring)

⬭ none ⬭ collarlike

⬭ skirtlike ⬭ sheathlike

Volva

⬭ none ⬭ present

Other (bruising, bleeding, smell, etc.) ______

______

Mushroom Identification Log

Location

Site ______________________ **Date** __________

◯ living tree ◯ leaf litter ◯ mulch ◯ other mushrooms
◯ dead tree or wood ◯ grass ◯ soil ◯ other __________

Type of tree(s) on or near __________

Forest type __________

Other (weather, lighting, humidity, etc.): __________

General

Size (cap, stem, & overall height): __________

Texture: tough, brittle, leathery, woody, soft, slimy, spongy, powdery, waxy, rubbery, watery, other: __________

__________ Color __________

Spore Color __________

Cap Shape & Features

bell-shaped

conical

convex

cylindrical

depressed

flat

funnel-shaped

umbilicate

umbonate

Surface texture (warts, scales, slime, etc.) __________

Cap margin: smooth, inrolled, sinuous/wavy, other:

____

Color changes __________

Gill Features

◯ gills ◯ false gills ◯ pores ◯ teeth

Spacing

crowded

close

distant

Gill Attachment

adnexed
(narrowly attached)

adnate
(broadly attached)

decurrent
(running down the stalk)

free
(gills don't meet the stem)

sinuate
(notched)

sketch

Stem Shape

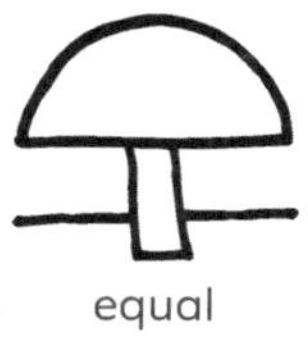

equal

club-shaped

bulbous

with cup
(volva)

tapering
downward

tapering
upward

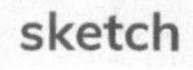

sketch

Other Features

Stem attachment ◯ central ◯ offset ◯ lateral

Stem texture: fleshy, fibrous, hollow, solid, fibrillose, scaly, velvety, smooth, other: ____________________

Veil ◯ none ◯ partial ◯ universal

Veil texture: thin, cobwebby, fibrillose, slimy, other: __________
____

Annulus (ring)

◯ none ◯ collarlike

◯ skirtlike ◯ sheathlike

Volva

◯ none ◯ present

Other (bruising, bleeding, smell, etc.) ____________________
____

Mushroom Identification Log

Location

Site ______________________ **Date** __________

◯ living tree ◯ leaf litter ◯ mulch ◯ other mushrooms
◯ dead tree or wood ◯ grass ◯ soil ◯ other __________

Type of tree(s) on or near ______________________

Forest type ______________________

Other (weather, lighting, humidity, etc.): __________

____

General

Size (cap, stem, & overall height): __________

Texture: tough, brittle, leathery, woody, soft, slimy, spongy, powdery, waxy, rubbery, watery, other: __________

______________________ Color __________

Spore Color ______________________

Cap Shape & Features

bell-shaped

conical

convex

cylindrical

depressed

flat

funnel-shaped

umbilicate

umbonate

Surface texture (warts, scales, slime, etc.) __________

____

Cap margin: smooth, inrolled, sinuous/wavy, other:

____

Color changes ______________________

Gill Features

◯ gills ◯ false gills ◯ pores ◯ teeth

Spacing

crowded

close

distant

Gill Attachment

adnexed
(narrowly attached)

adnate
(broadly attached)

decurrent
(running down the stalk)

free
(gills don't meet the stem)

sinuate
(notched)

sketch

Stem Shape

equal

club-shaped

bulbous

with cup
(volva)

tapering
downward

tapering
upward

sketch

Other Features

Stem attachment ◯ central ◯ offset ◯ lateral

Stem texture: fleshy, fibrous, hollow, solid, fibrillose, scaly, velvety, smooth, other: ______________________

Veil ◯ none ◯ partial ◯ universal

Veil texture: thin, cobwebby, fibrillose, slimy, other: ______________________

____

Annulus (ring)

◯ none ◯ collarlike

◯ skirtlike ◯ sheathlike

Volva

◯ none ◯ present

Other (bruising, bleeding, smell, etc.) ______________________

____

Mushroom Identification Log

Location

Site ______________________ **Date** ______________

◯ living tree ◯ leaf litter ◯ mulch ◯ other mushrooms
◯ dead tree or wood ◯ grass ◯ soil ◯ other______

Type of tree(s) on or near______________________

Forest type______________________

Other (weather, lighting, humidity, etc.): ______________________

____

General

Size (cap, stem, & overall height): ______________________

Texture: tough, brittle, leathery, woody, soft, slimy, spongy, powdery, waxy, rubbery, watery, other: ______________

______________________ Color ______________

Spore Color______________________

Cap Shape & Features

bell-shaped
conical
convex
cylindrical
depressed
flat
funnel-shaped
umbilicate
umbonate

Surface texture (warts, scales, slime, etc.) ______________________

____

Cap margin: smooth, inrolled, sinuous/wavy, other:

____

Color changes______________________

Gill Features

◯ gills ◯ false gills ◯ pores ◯ teeth

Spacing

crowded
close
distant

Gill Attachment

adnexed
(narrowly attached)

adnate
(broadly attached)

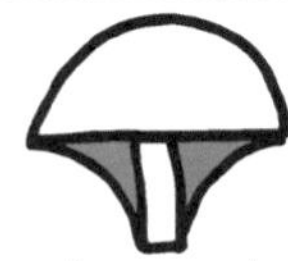

decurrent
(running down the stalk)

free
(gills don't meet the stem)

sinuate
(notched)

sketch

Stem Shape

equal

club-shaped

bulbous

with cup
(volva)

tapering
downward

tapering
upward

sketch

Other Features

Stem attachment ◯ central ◯ offset ◯ lateral

Stem texture: fleshy, fibrous, hollow, solid, fibrillose, scaly, velvety, smooth, other: ______________________

Veil ◯ none ◯ partial ◯ universal

Veil texture: thin, cobwebby, fibrillose, slimy, other: ________

____

Annulus (ring)

◯ none ◯ collarlike

◯ skirtlike ◯ sheathlike

Volva

◯ none ◯ present

Other (bruising, bleeding, smell, etc.) ______________________

____

Mushroom Identification Log

Location

Site ______________________ **Date** __________

◯ living tree ◯ leaf litter ◯ mulch ◯ other mushrooms
◯ dead tree or wood ◯ grass ◯ soil ◯ other__________

Type of tree(s) on or near____________________

Forest type____________________

Other (weather, lighting, humidity, etc.): ____________________

General

Size (cap, stem, & overall height): ____________________

Texture: tough, brittle, leathery, woody, soft, slimy, spongy, powdery, waxy, rubbery, watery, other: ____________________

____________________ Color __________

Spore Color____________________

Cap Shape & Features

bell-shaped

conical

convex

cylindrical

depressed

flat

funnel-shaped

umbilicate

umbonate

Surface texture (warts, scales, slime, etc.) ____________________

Cap margin: smooth, inrolled, sinuous/wavy, other:

____

Color changes____________________

Gill Features

◯ gills ◯ false gills ◯ pores ◯ teeth

Spacing

crowded

close

distant

Gill Attachment

adnexed
(narrowly attached)

adnate
(broadly attached)

decurrent
(running down the stalk)

free
(gills don't meet the stem)

sinuate
(notched)

Stem Shape

equal

club-shaped

bulbous

with cup
(volva)

tapering
downward

tapering
upward

sketch

Other Features

Stem attachment ◯ central ◯ offset ◯ lateral

Stem texture: fleshy, fibrous, hollow, solid, fibrillose, scaly, velvety, smooth, other: ____________________

Veil ◯ none ◯ partial ◯ universal

Veil texture: thin, cobwebby, fibrillose, slimy, other: __________
____

Annulus (ring)
◯ none ◯ collarlike
◯ skirtlike ◯ sheathlike

Volva
◯ none ◯ present

Other (bruising, bleeding, smell, etc.) ____________________
____

Mushroom Identification Log

Location

Site ______________________ **Date** __________

◯ living tree ◯ leaf litter ◯ mulch ◯ other mushrooms
◯ dead tree or wood ◯ grass ◯ soil ◯ other__________

Type of tree(s) on or near____________________

Forest type____________________

Other (weather, lighting, humidity, etc.): ____________________

____

General

Size (cap, stem, & overall height): ____________________

Texture: tough, brittle, leathery, woody, soft, slimy, spongy, powdery, waxy, rubbery, watery, other: ____________________

____________________ Color __________

Spore Color____________________

Cap Shape & Features

bell-shaped
conical
convex
cylindrical
depressed
flat
funnel-shaped
umbilicate
umbonate

Surface texture (warts, scales, slime, etc.) ____________________

____

Cap margin: smooth, inrolled, sinuous/wavy, other:

____

Color changes____________________

Gill Features

◯ gills ◯ false gills ◯ pores ◯ teeth

Spacing

crowded
close
distant

Gill Attachment

adnexed
(narrowly attached)

adnate
(broadly attached)

decurrent
(running down the stalk)

free
(gills don't meet the stem)

sinuate
(notched)

Stem Shape

equal

club-shaped

bulbous

with cup
(volva)

tapering
downward

tapering
upward

Other Features

Stem attachment ◯ central ◯ offset ◯ lateral

Stem texture: fleshy, fibrous, hollow, solid, fibrillose, scaly, velvety, smooth, other: ______________________

Veil ◯ none ◯ partial ◯ universal

Veil texture: thin, cobwebby, fibrillose, slimy, other: ______________________

Annulus (ring)

◯ none ◯ collarlike
◯ skirtlike ◯ sheathlike

Volva

◯ none ◯ present

Other (bruising, bleeding, smell, etc.) ______________________

Mushroom Identification Log

Location

Site ________________ **Date** ________

◯ living tree ◯ leaf litter ◯ mulch ◯ other mushrooms
◯ dead tree or wood ◯ grass ◯ soil ◯ other ________

Type of tree(s) on or near ________________

Forest type ________________

Other (weather, lighting, humidity, etc.): ________________

____

General

Size (cap, stem, & overall height): ________________

Texture: tough, brittle, leathery, woody, soft, slimy, spongy, powdery, waxy, rubbery, watery, other: ________________

________________ Color ________

Spore Color ________________

Cap Shape & Features

bell-shaped
conical
convex
cylindrical
depressed
flat
funnel-shaped
umbilicate
umbonate

Surface texture (warts, scales, slime, etc.) ________________

____

Cap margin: smooth, inrolled, sinuous/wavy, other:

____

Color changes ________________

Gill Features

◯ gills ◯ false gills ◯ pores ◯ teeth

Spacing

crowded
close
distant

Gill Attachment

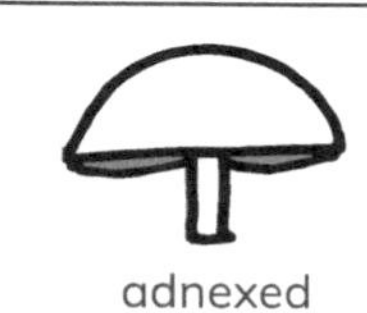
adnexed
(narrowly attached)

adnate
(broadly attached)

decurrent
(running down the stalk)

free
(gills don't meet the stem)

sinuate
(notched)

sketch

Stem Shape

equal

club-shaped

bulbous

with cup
(volva)

tapering
downward

tapering
upward

sketch

Other Features

Stem attachment ◯ central ◯ offset ◯ lateral

Stem texture: fleshy, fibrous, hollow, solid, fibrillose, scaly, velvety, smooth, other: ____________

Veil ◯ none ◯ partial ◯ universal

Veil texture: thin, cobwebby, fibrillose, slimy, other: ____________

____

Annulus (ring)
◯ none ◯ collarlike
◯ skirtlike ◯ sheathlike

Volva
◯ none ◯ present

Other (bruising, bleeding, smell, etc.) ____________

____

Mushroom Identification Log

Location

Site ______ **Date** ______

◯ living tree ◯ leaf litter ◯ mulch ◯ other mushrooms
◯ dead tree or wood ◯ grass ◯ soil ◯ other ______

Type of tree(s) on or near ______

Forest type ______

Other (weather, lighting, humidity, etc.): ______

______

General

Size (cap, stem, & overall height): ______

Texture: tough, brittle, leathery, woody, soft, slimy, spongy, powdery, waxy, rubbery, watery, other: ______

______ Color ______

Spore Color ______

Cap Shape & Features

bell-shaped
conical
convex
cylindrical
depressed

flat
funnel-shaped
umbilicate
umbonate

Surface texture (warts, scales, slime, etc.) ______

______

Cap margin: smooth, inrolled, sinuous/wavy, other:

______

Color changes ______

Gill Features

◯ gills ◯ false gills ◯ pores ◯ teeth

Spacing

crowded
close
distant

Gill Attachment

adnexed
(narrowly attached)

adnate
(broadly attached)

decurrent
(running down the stalk)

free
(gills don't meet the stem)

sinuate
(notched)

sketch

Stem Shape

equal

club-shaped

bulbous

with cup
(volva)

tapering
downward

tapering
upward

sketch

Other Features

Stem attachment ◯ central ◯ offset ◯ lateral

Stem texture: fleshy, fibrous, hollow, solid, fibrillose, scaly, velvety, smooth, other: ____________________

Veil ◯ none ◯ partial ◯ universal

Veil texture: thin, cobwebby, fibrillose, slimy, other: ________

____

Annulus (ring)
◯ none ◯ collarlike
◯ skirtlike ◯ sheathlike

Volva
◯ none ◯ present

Other (bruising, bleeding, smell, etc.) ____________________

____

Mushroom Identification Log

Location

Site ______ **Date** ______

◯ living tree ◯ leaf litter ◯ mulch ◯ other mushrooms
◯ dead tree or wood ◯ grass ◯ soil ◯ other ______

Type of tree(s) on or near ______

Forest type ______

Other (weather, lighting, humidity, etc.): ______

______

General

Size (cap, stem, & overall height): ______

Texture: tough, brittle, leathery, woody, soft, slimy, spongy, powdery, waxy, rubbery, watery, other: ______

______ Color ______

Spore Color ______

Cap Shape & Features

bell-shaped, conical, convex, cylindrical, depressed

flat, funnel-shaped, umbilicate, umbonate

Surface texture (warts, scales, slime, etc.) ______

______

Cap margin: smooth, inrolled, sinuous/wavy, other:

______

Color changes ______

Gill Features

◯ gills ◯ false gills ◯ pores ◯ teeth

Spacing

crowded, close, distant

Gill Attachment

adnexed (narrowly attached)

adnate (broadly attached)

decurrent (running down the stalk)

free (gills don't meet the stem)

sinuate (notched)

Stem Shape

equal

club-shaped

bulbous

with cup (volva)

tapering downward

tapering upward

Other Features

Stem attachment ◯ central ◯ offset ◯ lateral

Stem texture: fleshy, fibrous, hollow, solid, fibrillose, scaly, velvety, smooth, other: ____________________

Veil ◯ none ◯ partial ◯ universal

Veil texture: thin, cobwebby, fibrillose, slimy, other: ________

____

Annulus (ring)

◯ none ◯ collarlike

◯ skirtlike ◯ sheathlike

Volva

◯ none ◯ present

Other (bruising, bleeding, smell, etc.) ____________________

____

Mushroom Identification Log

Location

Site ______________________ **Date** __________

◯ living tree ◯ leaf litter ◯ mulch ◯ other mushrooms
◯ dead tree or wood ◯ grass ◯ soil ◯ other__________

Type of tree(s) on or near______________________

Forest type______________________

Other (weather, lighting, humidity, etc.): __________

____

General

Size (cap, stem, & overall height): __________

Texture: tough, brittle, leathery, woody, soft, slimy, spongy, powdery, waxy, rubbery, watery, other: __________

______________________ Color __________

Spore Color______________________

Cap Shape & Features

Surface texture (warts, scales, slime, etc.) __________

____

Cap margin: smooth, inrolled, sinuous/wavy, other:

____

Color changes______________________

Gill Features

◯ gills ◯ false gills ◯ pores ◯ teeth

Spacing

crowded

close

distant

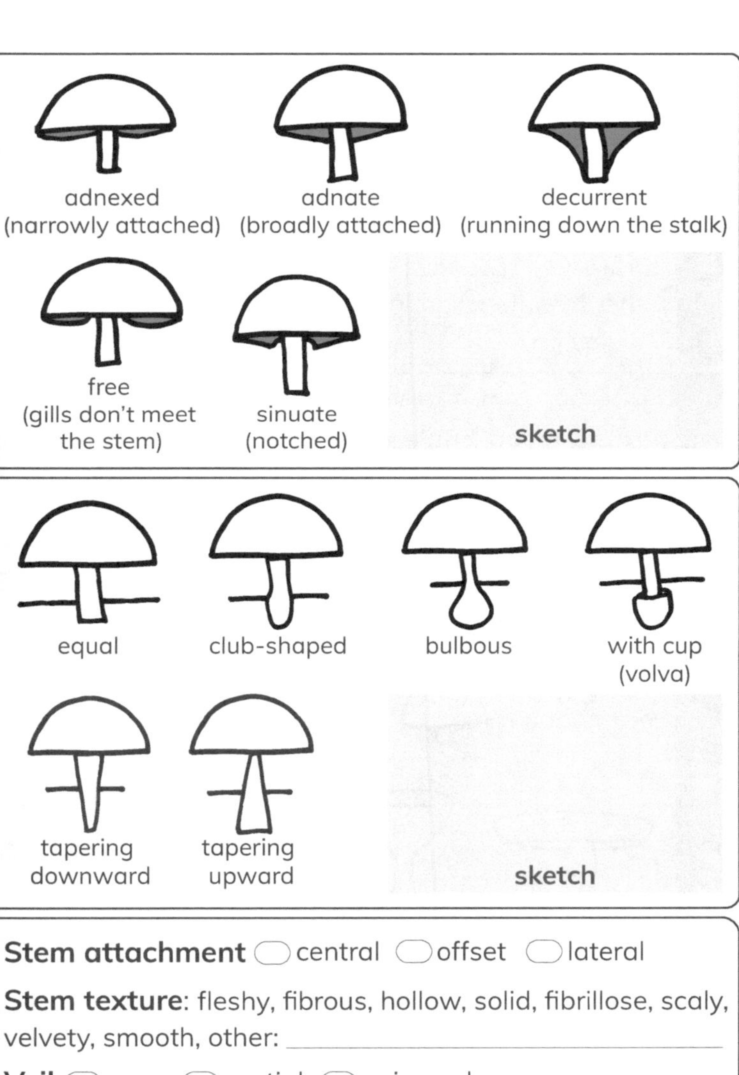

Other Features

Stem attachment ◯ central ◯ offset ◯ lateral

Stem texture: fleshy, fibrous, hollow, solid, fibrillose, scaly, velvety, smooth, other: ______________________

Veil ◯ none ◯ partial ◯ universal

Veil texture: thin, cobwebby, fibrillose, slimy, other: ______________________

____

Annulus (ring)

◯ none ◯ collarlike

◯ skirtlike ◯ sheathlike

Volva

◯ none ◯ present

Other (bruising, bleeding, smell, etc.) ______________________

____

Mushroom Identification Log

Location

Site ______________________ **Date** ______

◯ living tree ◯ leaf litter ◯ mulch ◯ other mushrooms
◯ dead tree or wood ◯ grass ◯ soil ◯ other ______

Type of tree(s) on or near ______

Forest type ______

Other (weather, lighting, humidity, etc.): ______

______

General

Size (cap, stem, & overall height): ______

Texture: tough, brittle, leathery, woody, soft, slimy, spongy, powdery, waxy, rubbery, watery, other: ______

______ Color ______

Spore Color ______

Cap Shape & Features

bell-shaped
conical
convex
cylindrical
depressed
flat
funnel-shaped
umbilicate
umbonate

Surface texture (warts, scales, slime, etc.) ______

______

Cap margin: smooth, inrolled, sinuous/wavy, other:

______

Color changes ______

Gill Features

◯ gills ◯ false gills ◯ pores ◯ teeth

Spacing

crowded
close
distant

Gill Attachment

adnexed (narrowly attached)

adnate (broadly attached)

decurrent (running down the stalk)

free (gills don't meet the stem)

sinuate (notched)

sketch

Stem Shape

equal

club-shaped

bulbous

with cup (volva)

tapering downward

tapering upward

sketch

Other Features

Stem attachment ◯ central ◯ offset ◯ lateral

Stem texture: fleshy, fibrous, hollow, solid, fibrillose, scaly, velvety, smooth, other: ______________________

Veil ◯ none ◯ partial ◯ universal

Veil texture: thin, cobwebby, fibrillose, slimy, other: __________

____

Annulus (ring)

◯ none ◯ collarlike

◯ skirtlike ◯ sheathlike

Volva

◯ none ◯ present

Other (bruising, bleeding, smell, etc.) ______________________

____

Mushroom Identification Log

Location

Site __________ **Date** __________

◯ living tree ◯ leaf litter ◯ mulch ◯ other mushrooms
◯ dead tree or wood ◯ grass ◯ soil ◯ other __________

Type of tree(s) on or near __________

Forest type __________

Other (weather, lighting, humidity, etc.): __________

General

Size (cap, stem, & overall height): __________

Texture: tough, brittle, leathery, woody, soft, slimy, spongy, powdery, waxy, rubbery, watery, other: __________

Color __________

Spore Color __________

Cap Shape & Features

bell-shaped
conical
convex
cylindrical
depressed
flat
funnel-shaped
umbilicate
umbonate

Surface texture (warts, scales, slime, etc.) __________

Cap margin: smooth, inrolled, sinuous/wavy, other: __________

Color changes __________

Gill Features

◯ gills ◯ false gills ◯ pores ◯ teeth

Spacing

crowded
close
distant

Gill Attachment

adnexed (narrowly attached)

adnate (broadly attached)

decurrent (running down the stalk)

free (gills don't meet the stem)

sinuate (notched)

sketch

Stem Shape

equal

club-shaped

bulbous

with cup (volva)

tapering downward

tapering upward

sketch

Other Features

Stem attachment ◯ central ◯ offset ◯ lateral

Stem texture: fleshy, fibrous, hollow, solid, fibrillose, scaly, velvety, smooth, other: ____________

Veil ◯ none ◯ partial ◯ universal

Veil texture: thin, cobwebby, fibrillose, slimy, other: ____________

Annulus (ring)

◯ none ◯ collarlike

◯ skirtlike ◯ sheathlike

Volva

◯ none ◯ present

Other (bruising, bleeding, smell, etc.) ____________

Mushroom Identification Log

Location

Site ______________________ **Date** ____________

◯ living tree ◯ leaf litter ◯ mulch ◯ other mushrooms
◯ dead tree or wood ◯ grass ◯ soil ◯ other ____________

Type of tree(s) on or near ______________________

Forest type ______________________

Other (weather, lighting, humidity, etc.): ______________________

____

General

Size (cap, stem, & overall height): ______________________

Texture: tough, brittle, leathery, woody, soft, slimy, spongy, powdery, waxy, rubbery, watery, other: ____________

______________________ Color ____________

Spore Color ______________________

Cap Shape & Features

bell-shaped, conical, convex, cylindrical, depressed

flat, funnel-shaped, umbilicate, umbonate

Surface texture (warts, scales, slime, etc.) ____________

____

Cap margin: smooth, inrolled, sinuous/wavy, other:

____

Color changes ______________________

Gill Features

◯ gills ◯ false gills ◯ pores ◯ teeth

Spacing

crowded, close, distant

Gill Attachment

adnexed
(narrowly attached)

adnate
(broadly attached)

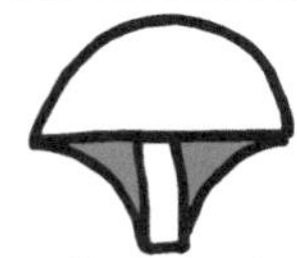

decurrent
(running down the stalk)

free
(gills don't meet the stem)

sinuate
(notched)

sketch

Stem Shape

equal

club-shaped

bulbous

with cup
(volva)

tapering downward

tapering upward

sketch

Other Features

Stem attachment ◯ central ◯ offset ◯ lateral

Stem texture: fleshy, fibrous, hollow, solid, fibrillose, scaly, velvety, smooth, other: ______

Veil ◯ none ◯ partial ◯ universal

Veil texture: thin, cobwebby, fibrillose, slimy, other: ______

______

Annulus (ring)

◯ none ◯ collarlike
◯ skirtlike ◯ sheathlike

Volva

◯ none ◯ present

Other (bruising, bleeding, smell, etc.) ______

______

Mushroom Identification Log

Location

Site ________________________ **Date** ____________

◯ living tree ◯ leaf litter ◯ mulch ◯ other mushrooms
◯ dead tree or wood ◯ grass ◯ soil ◯ other ________

Type of tree(s) on or near ____________________

Forest type ____________________

Other (weather, lighting, humidity, etc.): ____________________

____

General

Size (cap, stem, & overall height): ____________________

Texture: tough, brittle, leathery, woody, soft, slimy, spongy, powdery, waxy, rubbery, watery, other: ____________________

____________________ Color ____________

Spore Color ____________________

Cap Shape & Features

bell-shaped

conical

convex

cylindrical

depressed

flat

funnel-shaped

umbilicate

umbonate

Surface texture (warts, scales, slime, etc.) ____________________

____

Cap margin: smooth, inrolled, sinuous/wavy, other:

____

Color changes ____________________

Gill Features

◯ gills ◯ false gills ◯ pores ◯ teeth

Spacing

crowded

close

distant

Gill Attachment

adnexed
(narrowly attached)

adnate
(broadly attached)

decurrent
(running down the stalk)

free
(gills don't meet the stem)

sinuate
(notched)

Stem Shape

equal

club-shaped

bulbous

with cup
(volva)

tapering
downward

tapering
upward

sketch

Other Features

Stem attachment ◯ central ◯ offset ◯ lateral

Stem texture: fleshy, fibrous, hollow, solid, fibrillose, scaly, velvety, smooth, other: ____________________

Veil ◯ none ◯ partial ◯ universal

Veil texture: thin, cobwebby, fibrillose, slimy, other: ____________________

____

Annulus (ring)

◯ none ◯ collarlike
◯ skirtlike ◯ sheathlike

Volva

◯ none ◯ present

Other (bruising, bleeding, smell, etc.) ____________________

____

Mushroom Identification Log

Location

Site ____________________ **Date** __________

◯ living tree ◯ leaf litter ◯ mulch ◯ other mushrooms
◯ dead tree or wood ◯ grass ◯ soil ◯ other __________

Type of tree(s) on or near ____________________

Forest type ____________________

Other (weather, lighting, humidity, etc.): ____________________

____

General

Size (cap, stem, & overall height): ____________________

Texture: tough, brittle, leathery, woody, soft, slimy, spongy, powdery, waxy, rubbery, watery, other: __________

____________________ Color __________

Spore Color ____________________

Cap Shape & Features

bell-shaped

conical

convex

cylindrical

depressed

flat

funnel-shaped

umbilicate

umbonate

Surface texture (warts, scales, slime, etc.) ____________________

____

Cap margin: smooth, inrolled, sinuous/wavy, other:

____

Color changes ____________________

Gill Features

◯ gills ◯ false gills ◯ pores ◯ teeth

Spacing

crowded

close

distant

Gill Attachment

adnexed
(narrowly attached)

adnate
(broadly attached)

decurrent
(running down the stalk)

free
(gills don't meet the stem)

sinuate
(notched)

sketch

Stem Shape

equal

club-shaped

bulbous

with cup
(volva)

tapering
downward

tapering
upward

sketch

Other Features

Stem attachment ◯ central ◯ offset ◯ lateral

Stem texture: fleshy, fibrous, hollow, solid, fibrillose, scaly, velvety, smooth, other: ______________________

Veil ◯ none ◯ partial ◯ universal

Veil texture: thin, cobwebby, fibrillose, slimy, other: ______

____

Annulus (ring)

◯ none ◯ collarlike
◯ skirtlike ◯ sheathlike

Volva

◯ none ◯ present

Other (bruising, bleeding, smell, etc.) ______________

____

Mushroom Identification Log

Location

Site ______________________ **Date** ____________

◯ living tree ◯ leaf litter ◯ mulch ◯ other mushrooms
◯ dead tree or wood ◯ grass ◯ soil ◯ other ________

Type of tree(s) on or near ______________________

Forest type ______________________

Other (weather, lighting, humidity, etc.): ______________________

____

General

Size (cap, stem, & overall height): ______________________

Texture: tough, brittle, leathery, woody, soft, slimy, spongy, powdery, waxy, rubbery, watery, other: ______________________

______________________ Color ____________

Spore Color ______________________

Cap Shape & Features

bell-shaped, conical, convex, cylindrical, depressed

flat, funnel-shaped, umbilicate, umbonate

Surface texture (warts, scales, slime, etc.) ______________________

____

Cap margin: smooth, inrolled, sinuous/wavy, other:

____

Color changes ______________________

Gill Features

◯ gills ◯ false gills ◯ pores ◯ teeth

Spacing

crowded, close, distant

Gill Attachment

adnexed
(narrowly attached)

adnate
(broadly attached)

decurrent
(running down the stalk)

free
(gills don't meet
the stem)

sinuate
(notched)

sketch

Stem Shape

equal

club-shaped

bulbous

with cup
(volva)

tapering
downward

tapering
upward

sketch

Other Features

Stem attachment ◯ central ◯ offset ◯ lateral

Stem texture: fleshy, fibrous, hollow, solid, fibrillose, scaly, velvety, smooth, other: ____________________

Veil ◯ none ◯ partial ◯ universal

Veil texture: thin, cobwebby, fibrillose, slimy, other: ____________________

____

Annulus (ring)

◯ none ◯ collarlike
◯ skirtlike ◯ sheathlike

Volva

◯ none ◯ present

Other (bruising, bleeding, smell, etc.) ____________________

____

Mushroom Identification Log

Location

Site ______________________ **Date** ______________

◯ living tree ◯ leaf litter ◯ mulch ◯ other mushrooms
◯ dead tree or wood ◯ grass ◯ soil ◯ other ______________

Type of tree(s) on or near ______________________

Forest type ______________________

Other (weather, lighting, humidity, etc.): ______________________

____

General

Size (cap, stem, & overall height): ______________________

Texture: tough, brittle, leathery, woody, soft, slimy, spongy, powdery, waxy, rubbery, watery, other: ______________

______________________ Color ______________

Spore Color ______________________

Cap Shape & Features

bell-shaped

conical

convex

cylindrical

depressed

flat

funnel-shaped

umbilicate

umbonate

Surface texture (warts, scales, slime, etc.) ______________

____

Cap margin: smooth, inrolled, sinuous/wavy, other:

____

Color changes ______________________

Gill Features

◯ gills ◯ false gills ◯ pores ◯ teeth

Spacing

crowded

close

distant

Gill Attachment

adnexed
(narrowly attached)

adnate
(broadly attached)

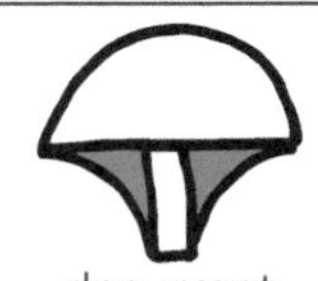
decurrent
(running down the stalk)

free
(gills don't meet the stem)

sinuate
(notched)

sketch

Stem Shape

equal

club-shaped

bulbous

with cup
(volva)

tapering
downward

tapering
upward

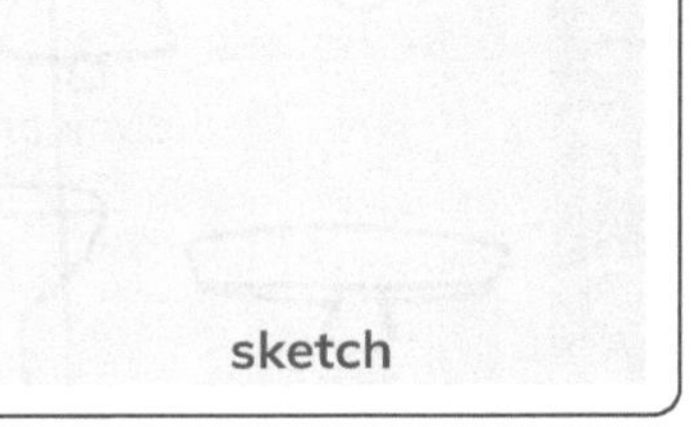
sketch

Other Features

Stem attachment ◯ central ◯ offset ◯ lateral

Stem texture: fleshy, fibrous, hollow, solid, fibrillose, scaly, velvety, smooth, other: ______

Veil ◯ none ◯ partial ◯ universal

Veil texture: thin, cobwebby, fibrillose, slimy, other: ______

______

Annulus (ring)

◯ none ◯ collarlike

◯ skirtlike ◯ sheathlike

Volva

◯ none ◯ present

Other (bruising, bleeding, smell, etc.) ______

______

Mushroom Identification Log

Location

Site ________________ **Date** ________

◯ living tree ◯ leaf litter ◯ mulch ◯ other mushrooms
◯ dead tree or wood ◯ grass ◯ soil ◯ other ________

Type of tree(s) on or near ________________

Forest type ________________

Other (weather, lighting, humidity, etc.): ________________

____

General

Size (cap, stem, & overall height): ________________

Texture: tough, brittle, leathery, woody, soft, slimy, spongy, powdery, waxy, rubbery, watery, other: ________

________________ Color ________

Spore Color ________________

Cap Shape & Features

bell-shaped

conical

convex

cylindrical

depressed

flat

funnel-shaped

umbilicate

umbonate

Surface texture (warts, scales, slime, etc.) ________________

____

Cap margin: smooth, inrolled, sinuous/wavy, other:

____

Color changes ________________

Gill Features

◯ gills ◯ false gills ◯ pores ◯ teeth

Spacing

crowded

close

distant

Gill Attachment

adnexed
(narrowly attached)

adnate
(broadly attached)

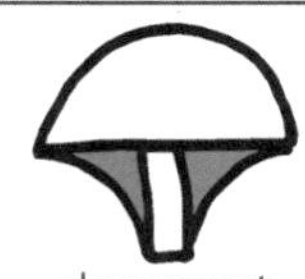

decurrent
(running down the stalk)

free
(gills don't meet the stem)

sinuate
(notched)

sketch

Stem Shape

equal

club-shaped

bulbous

with cup
(volva)

tapering
downward

tapering
upward

sketch

Other Features

Stem attachment ◯ central ◯ offset ◯ lateral

Stem texture: fleshy, fibrous, hollow, solid, fibrillose, scaly, velvety, smooth, other: ____________________

Veil ◯ none ◯ partial ◯ universal

Veil texture: thin, cobwebby, fibrillose, slimy, other: ____________

____

Annulus (ring)

◯ none ◯ collarlike
◯ skirtlike ◯ sheathlike

Volva

◯ none ◯ present

Other (bruising, bleeding, smell, etc.) ____________________

____

Mushroom Identification Log

Location

Site ______________________ Date ____________

◯ living tree ◯ leaf litter ◯ mulch ◯ other mushrooms
◯ dead tree or wood ◯ grass ◯ soil ◯ other ________

Type of tree(s) on or near ______________________

Forest type ______________________

Other (weather, lighting, humidity, etc.): ______________________

____

General

Size (cap, stem, & overall height): ______________________

Texture: tough, brittle, leathery, woody, soft, slimy, spongy, powdery, waxy, rubbery, watery, other: ______________________

______________________ Color ____________

Spore Color ______________________

Cap Shape & Features

bell-shaped
conical
convex
cylindrical
depressed
flat
funnel-shaped
umbilicate
umbonate

Surface texture (warts, scales, slime, etc.) ______________________

____

Cap margin: smooth, inrolled, sinuous/wavy, other:

____

Color changes ______________________

Gill Features

◯ gills ◯ false gills ◯ pores ◯ teeth

Spacing

crowded
close
distant

Gill Attachment

adnexed
(narrowly attached)

adnate
(broadly attached)

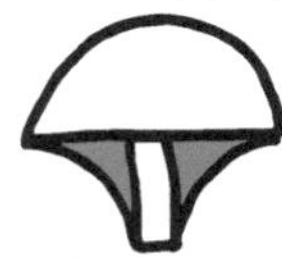

decurrent
(running down the stalk)

free
(gills don't meet the stem)

sinuate
(notched)

sketch

Stem Shape

equal

club-shaped

bulbous

with cup
(volva)

tapering downward

tapering upward

sketch

Other Features

Stem attachment ◯ central ◯ offset ◯ lateral

Stem texture: fleshy, fibrous, hollow, solid, fibrillose, scaly, velvety, smooth, other: ____________

Veil ◯ none ◯ partial ◯ universal

Veil texture: thin, cobwebby, fibrillose, slimy, other: ____________

____

Annulus (ring)

◯ none ◯ collarlike

◯ skirtlike ◯ sheathlike

Volva

◯ none ◯ present

Other (bruising, bleeding, smell, etc.) ____________

____

Mushroom Identification Log

Location

Site ______________________ **Date** __________

◯ living tree ◯ leaf litter ◯ mulch ◯ other mushrooms
◯ dead tree or wood ◯ grass ◯ soil ◯ other ________

Type of tree(s) on or near ____________________

Forest type ____________________

Other (weather, lighting, humidity, etc.): ____________________

____

General

Size (cap, stem, & overall height): ____________________

Texture: tough, brittle, leathery, woody, soft, slimy, spongy, powdery, waxy, rubbery, watery, other: ____________________

____________________ Color ____________________

Spore Color ____________________

Cap Shape & Features

bell-shaped

conical

convex

cylindrical

depressed

flat

funnel-shaped

umbilicate

umbonate

Surface texture (warts, scales, slime, etc.) ____________________

____

Cap margin: smooth, inrolled, sinuous/wavy, other:

____

Color changes ____________________

Gill Features

◯ gills ◯ false gills ◯ pores ◯ teeth

Spacing

crowded

close

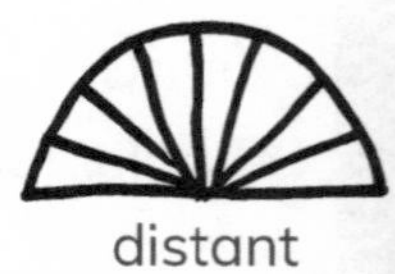

distant

Gill Attachment

adnexed
(narrowly attached)

adnate
(broadly attached)

decurrent
(running down the stalk)

free
(gills don't meet the stem)

sinuate
(notched)

sketch

Stem Shape

equal

club-shaped

bulbous

with cup
(volva)

tapering
downward

tapering
upward

sketch

Other Features

Stem attachment ◯ central ◯ offset ◯ lateral

Stem texture: fleshy, fibrous, hollow, solid, fibrillose, scaly, velvety, smooth, other: ______________________________

Veil ◯ none ◯ partial ◯ universal

Veil texture: thin, cobwebby, fibrillose, slimy, other: ______

____

Annulus (ring)

◯ none ◯ collarlike

◯ skirtlike ◯ sheathlike

Volva

◯ none ◯ present

Other (bruising, bleeding, smell, etc.) ____________________

____

Mushroom Identification Log

Location

Site ______________________ **Date** __________

◯ living tree ◯ leaf litter ◯ mulch ◯ other mushrooms

◯ dead tree or wood ◯ grass ◯ soil ◯ other __________

Type of tree(s) on or near ______________________

Forest type ______________________

Other (weather, lighting, humidity, etc.): ______________________

____

General

Size (cap, stem, & overall height): ______________________

Texture: tough, brittle, leathery, woody, soft, slimy, spongy, powdery, waxy, rubbery, watery, other: __________

______________________ Color __________

Spore Color ______________________

Cap Shape & Features

bell-shaped

conical

convex

cylindrical

depressed

flat

funnel-shaped

umbilicate

umbonate

Surface texture (warts, scales, slime, etc.) __________

____

Cap margin: smooth, inrolled, sinuous/wavy, other:

____

Color changes ______________________

Gill Features

◯ gills ◯ false gills ◯ pores ◯ teeth

Spacing

crowded

close

distant

Gill Attachment

adnexed
(narrowly attached)

adnate
(broadly attached)

decurrent
(running down the stalk)

free
(gills don't meet the stem)

sinuate
(notched)

sketch

Stem Shape

equal

club-shaped

bulbous

with cup
(volva)

tapering
downward

tapering
upward

sketch

Other Features

Stem attachment ◯ central ◯ offset ◯ lateral

Stem texture: fleshy, fibrous, hollow, solid, fibrillose, scaly, velvety, smooth, other: ______________________

Veil ◯ none ◯ partial ◯ universal

Veil texture: thin, cobwebby, fibrillose, slimy, other: __________

____

Annulus (ring)

◯ none ◯ collarlike

◯ skirtlike ◯ sheathlike

Volva

◯ none ◯ present

Other (bruising, bleeding, smell, etc.) ______________________

____

Mushroom Identification Log

Location

Site ______________________ **Date** __________

◯ living tree ◯ leaf litter ◯ mulch ◯ other mushrooms

◯ dead tree or wood ◯ grass ◯ soil ◯ other __________

Type of tree(s) on or near ______________________

Forest type ______________________

Other (weather, lighting, humidity, etc.): ______________________

____

General

Size (cap, stem, & overall height): ______________________

Texture: tough, brittle, leathery, woody, soft, slimy, spongy, powdery, waxy, rubbery, watery, other: __________

______________________ Color __________

Spore Color ______________________

Cap Shape & Features

bell-shaped

conical

convex

cylindrical

depressed

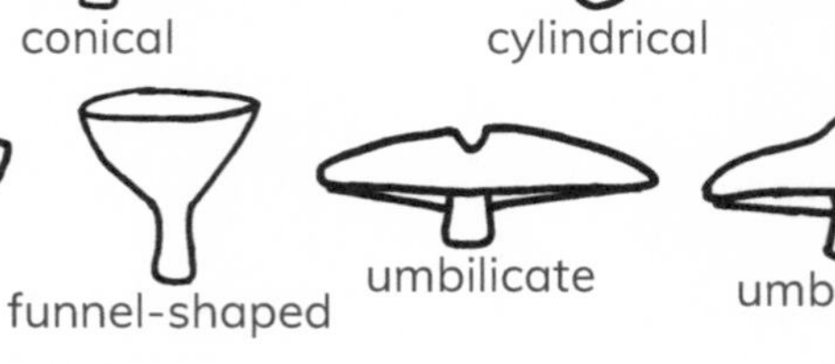

Surface texture (warts, scales, slime, etc.) ______________________

____

Cap margin: smooth, inrolled, sinuous/wavy, other:

____

Color changes ______________________

Gill Features

◯ gills ◯ false gills ◯ pores ◯ teeth

Spacing

crowded

close

distant

Gill Attachment

adnexed
(narrowly attached)

adnate
(broadly attached)

decurrent
(running down the stalk)

free
(gills don't meet the stem)

sinuate
(notched)

sketch

Stem Shape

equal

club-shaped

bulbous

with cup
(volva)

tapering
downward

tapering
upward

sketch

Other Features

Stem attachment ◯ central ◯ offset ◯ lateral

Stem texture: fleshy, fibrous, hollow, solid, fibrillose, scaly, velvety, smooth, other: ____________

Veil ◯ none ◯ partial ◯ universal

Veil texture: thin, cobwebby, fibrillose, slimy, other: ____________

Annulus (ring)

◯ none ◯ collarlike
◯ skirtlike ◯ sheathlike

Volva

◯ none ◯ present

Other (bruising, bleeding, smell, etc.) ____________

Mushroom Identification Log

Location

Site ______________________ **Date** __________

◯ living tree ◯ leaf litter ◯ mulch ◯ other mushrooms
◯ dead tree or wood ◯ grass ◯ soil ◯ other ________

Type of tree(s) on or near ______________________

Forest type ______________________

Other (weather, lighting, humidity, etc.): ______________________

____

General

Size (cap, stem, & overall height): ______________________

Texture: tough, brittle, leathery, woody, soft, slimy, spongy, powdery, waxy, rubbery, watery, other: ______________________

______________________ Color ______________________

Spore Color ______________________

Cap Shape & Features

bell-shaped, conical, convex, cylindrical, depressed

flat, funnel-shaped, umbilicate, umbonate

Surface texture (warts, scales, slime, etc.) ______________________

____

Cap margin: smooth, inrolled, sinuous/wavy, other:

____

Color changes ______________________

Gill Features

◯ gills ◯ false gills ◯ pores ◯ teeth

Spacing

crowded, close, distant

Gill Attachment

adnexed
(narrowly attached)

adnate
(broadly attached)

decurrent
(running down the stalk)

free
(gills don't meet the stem)

sinuate
(notched)

Stem Shape

equal

club-shaped

bulbous

with cup
(volva)

tapering
downward

tapering
upward

Other Features

Stem attachment ◯ central ◯ offset ◯ lateral

Stem texture: fleshy, fibrous, hollow, solid, fibrillose, scaly, velvety, smooth, other: ______________________

Veil ◯ none ◯ partial ◯ universal

Veil texture: thin, cobwebby, fibrillose, slimy, other: ______

____

Annulus (ring)
◯ none ◯ collarlike
◯ skirtlike ◯ sheathlike

Volva
◯ none ◯ present

Other (bruising, bleeding, smell, etc.) ______________

____

Mushroom Identification Log

Location

Site ______________________ **Date** __________

◯ living tree ◯ leaf litter ◯ mulch ◯ other mushrooms
◯ dead tree or wood ◯ grass ◯ soil ◯ other __________

Type of tree(s) on or near __________________________

Forest type ______________________________________

Other (weather, lighting, humidity, etc.): ______________

____

General

Size (cap, stem, & overall height): ___________________

Texture: tough, brittle, leathery, woody, soft, slimy, spongy, powdery, waxy, rubbery, watery, other: ______________

______________________________ Color ______________

Spore Color ______________________________________

Cap Shape & Features

bell-shaped

conical

convex

cylindrical

depressed

flat

funnel-shaped

umbilicate

umbonate

Surface texture (warts, scales, slime, etc.) ______________

____

Cap margin: smooth, inrolled, sinuous/wavy, other:

____

Color changes ______________________________________

Gill Features

◯ gills ◯ false gills ◯ pores ◯ teeth

Spacing

crowded

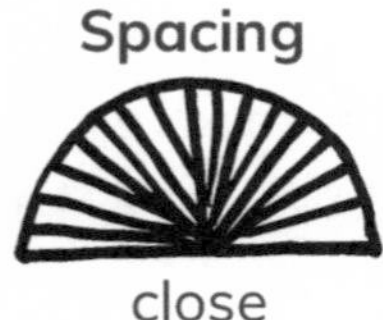

close

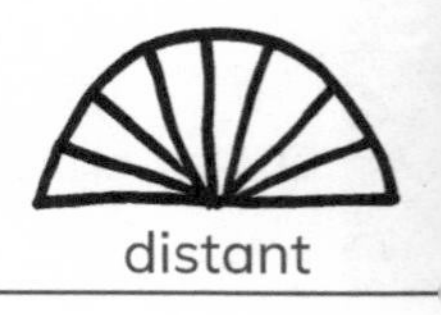

distant

Gill Attachment

adnexed
(narrowly attached)

adnate
(broadly attached)

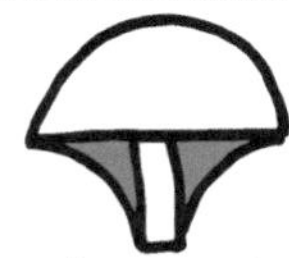

decurrent
(running down the stalk)

free
(gills don't meet
the stem)

sinuate
(notched)

sketch

Stem Shape

equal

club-shaped

bulbous

with cup
(volva)

tapering
downward

tapering
upward

sketch

Other Features

Stem attachment ◯ central ◯ offset ◯ lateral

Stem texture: fleshy, fibrous, hollow, solid, fibrillose, scaly, velvety, smooth, other: ____________________

Veil ◯ none ◯ partial ◯ universal

Veil texture: thin, cobwebby, fibrillose, slimy, other: __________

____

Annulus (ring)

◯ none ◯ collarlike
◯ skirtlike ◯ sheathlike

Volva

◯ none ◯ present

Other (bruising, bleeding, smell, etc.) ____________________

____

Mushroom Identification Log

Location

Site ______________________ **Date** __________

◯ living tree ◯ leaf litter ◯ mulch ◯ other mushrooms
◯ dead tree or wood ◯ grass ◯ soil ◯ other __________

Type of tree(s) on or near ______________________

Forest type ______________________

Other (weather, lighting, humidity, etc.): ______________________

____

General

Size (cap, stem, & overall height): ______________________

Texture: tough, brittle, leathery, woody, soft, slimy, spongy, powdery, waxy, rubbery, watery, other: __________

______________________ Color __________

Spore Color ______________________

Cap Shape & Features

bell-shaped conical convex cylindrical depressed

flat funnel-shaped umbilicate umbonate

Surface texture (warts, scales, slime, etc.) ______________________

____

Cap margin: smooth, inrolled, sinuous/wavy, other:

____

Color changes ______________________

Gill Features

◯ gills ◯ false gills ◯ pores ◯ teeth

Spacing

crowded

close

distant

Gill Attachment

adnexed
(narrowly attached)

adnate
(broadly attached)

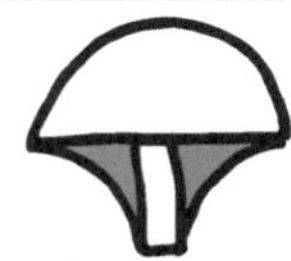
decurrent
(running down the stalk)

free
(gills don't meet the stem)

sinuate
(notched)

sketch

Stem Shape

equal

club-shaped

bulbous

with cup
(volva)

tapering
downward

tapering
upward

sketch

Other Features

Stem attachment ◯ central ◯ offset ◯ lateral

Stem texture: fleshy, fibrous, hollow, solid, fibrillose, scaly, velvety, smooth, other: ____________

Veil ◯ none ◯ partial ◯ universal

Veil texture: thin, cobwebby, fibrillose, slimy, other: ____________

____

Annulus (ring)
◯ none ◯ collarlike
◯ skirtlike ◯ sheathlike

Volva
◯ none ◯ present

Other (bruising, bleeding, smell, etc.) ____________

____

Mushroom Identification Log

Location

Site ____________________ **Date** __________

◯ living tree ◯ leaf litter ◯ mulch ◯ other mushrooms
◯ dead tree or wood ◯ grass ◯ soil ◯ other__________

Type of tree(s) on or near____________________

Forest type____________________

Other (weather, lighting, humidity, etc.): __________

____

General

Size (cap, stem, & overall height): __________

Texture: tough, brittle, leathery, woody, soft, slimy, spongy, powdery, waxy, rubbery, watery, other: __________

____________________ Color __________

Spore Color____________________

Cap Shape & Features

bell-shaped
conical
convex
cylindrical
depressed
flat
funnel-shaped
umbilicate
umbonate

Surface texture (warts, scales, slime, etc.) __________

____

Cap margin: smooth, inrolled, sinuous/wavy, other:

____

Color changes____________________

Gill Features

◯ gills ◯ false gills ◯ pores ◯ teeth

Spacing

crowded
close
distant

Gill Attachment

adnexed
(narrowly attached)

adnate
(broadly attached)

decurrent
(running down the stalk)

free
(gills don't meet the stem)

sinuate
(notched)

sketch

Stem Shape

equal

club-shaped

bulbous

with cup
(volva)

tapering
downward

tapering
upward

sketch

Other Features

Stem attachment ◯ central ◯ offset ◯ lateral

Stem texture: fleshy, fibrous, hollow, solid, fibrillose, scaly, velvety, smooth, other: ______

Veil ◯ none ◯ partial ◯ universal

Veil texture: thin, cobwebby, fibrillose, slimy, other: ______

______

Annulus (ring)

◯ none ◯ collarlike

◯ skirtlike ◯ sheathlike

Volva

◯ none ◯ present

Other (bruising, bleeding, smell, etc.) ______

______

Mushroom Identification Log

Location

Site ______ **Date** ______

◯ living tree ◯ leaf litter ◯ mulch ◯ other mushrooms
◯ dead tree or wood ◯ grass ◯ soil ◯ other ______

Type of tree(s) on or near ______

Forest type ______

Other (weather, lighting, humidity, etc.): ______

General

Size (cap, stem, & overall height): ______

Texture: tough, brittle, leathery, woody, soft, slimy, spongy, powdery, waxy, rubbery, watery, other: ______

______ Color ______

Spore Color ______

Cap Shape & Features

bell-shaped

conical

convex

cylindrical

depressed

flat

funnel-shaped

umbilicate

umbonate

Surface texture (warts, scales, slime, etc.) ______

Cap margin: smooth, inrolled, sinuous/wavy, other:

______

Color changes ______

Gill Features

◯ gills ◯ false gills ◯ pores ◯ teeth

Spacing

crowded

close

distant

Gill Attachment

adnexed
(narrowly attached)

adnate
(broadly attached)

decurrent
(running down the stalk)

free
(gills don't meet the stem)

sinuate
(notched)

sketch

Stem Shape

equal

club-shaped

bulbous

with cup
(volva)

tapering
downward

tapering
upward

sketch

Other Features

Stem attachment ◯ central ◯ offset ◯ lateral

Stem texture: fleshy, fibrous, hollow, solid, fibrillose, scaly, velvety, smooth, other: ______________________

Veil ◯ none ◯ partial ◯ universal

Veil texture: thin, cobwebby, fibrillose, slimy, other: __________

____

Annulus (ring)

◯ none ◯ collarlike

◯ skirtlike ◯ sheathlike

Volva

◯ none ◯ present

Other (bruising, bleeding, smell, etc.) ______________________

____

Mushroom Identification Log

Location

Site ______ **Date** ______

◯ living tree ◯ leaf litter ◯ mulch ◯ other mushrooms

◯ dead tree or wood ◯ grass ◯ soil ◯ other______

Type of tree(s) on or near______

Forest type______

Other (weather, lighting, humidity, etc.): ______

General

Size (cap, stem, & overall height): ______

Texture: tough, brittle, leathery, woody, soft, slimy, spongy, powdery, waxy, rubbery, watery, other: ______

______ Color______

Spore Color______

Cap Shape & Features

bell-shaped

conical

convex

cylindrical

depressed

flat

funnel-shaped

umbilicate

umbonate

Surface texture (warts, scales, slime, etc.) ______

Cap margin: smooth, inrolled, sinuous/wavy, other:

______

Color changes______

Gill Features

◯ gills ◯ false gills ◯ pores ◯ teeth

Spacing

crowded

close

distant

Gill Attachment

adnexed
(narrowly attached)

adnate
(broadly attached)

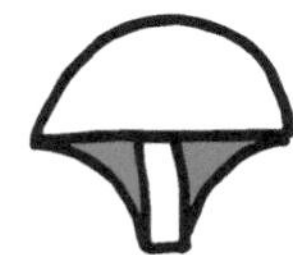
decurrent
(running down the stalk)

free
(gills don't meet the stem)

sinuate
(notched)

sketch

Stem Shape

equal

club-shaped

bulbous

with cup
(volva)

tapering
downward

tapering
upward

sketch

Other Features

Stem attachment ◯ central ◯ offset ◯ lateral

Stem texture: fleshy, fibrous, hollow, solid, fibrillose, scaly, velvety, smooth, other: ______

Veil ◯ none ◯ partial ◯ universal

Veil texture: thin, cobwebby, fibrillose, slimy, other: ______

______

Annulus (ring)

◯ none ◯ collarlike
◯ skirtlike ◯ sheathlike

Volva

◯ none ◯ present

Other (bruising, bleeding, smell, etc.) ______

______

Mushroom Identification Log

Location

Site ______________________ **Date** ______________________

◯ living tree ◯ leaf litter ◯ mulch ◯ other mushrooms
◯ dead tree or wood ◯ grass ◯ soil ◯ other ______

Type of tree(s) on or near ______________________

Forest type ______________________

Other (weather, lighting, humidity, etc.): ______________________

____

General

Size (cap, stem, & overall height): ______________________

Texture: tough, brittle, leathery, woody, soft, slimy, spongy, powdery, waxy, rubbery, watery, other: ______________________

______________________ Color ______________________

Spore Color ______________________

Cap Shape & Features

bell-shaped
conical
convex
cylindrical
depressed
flat
funnel-shaped
umbilicate
umbonate

Surface texture (warts, scales, slime, etc.) ______________________

____

Cap margin: smooth, inrolled, sinuous/wavy, other:

____

Color changes ______________________

Gill Features

◯ gills ◯ false gills ◯ pores ◯ teeth

Spacing

crowded
close
distant

Gill Attachment

adnexed
(narrowly attached)

adnate
(broadly attached)

decurrent
(running down the stalk)

free
(gills don't meet the stem)

sinuate
(notched)

sketch

Stem Shape

equal

club-shaped

bulbous

with cup
(volva)

tapering
downward

tapering
upward

sketch

Other Features

Stem attachment ◯ central ◯ offset ◯ lateral

Stem texture: fleshy, fibrous, hollow, solid, fibrillose, scaly, velvety, smooth, other: ______________________

Veil ◯ none ◯ partial ◯ universal

Veil texture: thin, cobwebby, fibrillose, slimy, other: ______

____

Annulus (ring)

◯ none ◯ collarlike

◯ skirtlike ◯ sheathlike

Volva

◯ none ◯ present

Other (bruising, bleeding, smell, etc.) ______________

____

Mushroom Identification Log

Location

Site ______________________ **Date** __________

◯ living tree ◯ leaf litter ◯ mulch ◯ other mushrooms
◯ dead tree or wood ◯ grass ◯ soil ◯ other ________

Type of tree(s) on or near ______________________

Forest type ______________________

Other (weather, lighting, humidity, etc.): ______________________

____

General

Size (cap, stem, & overall height): ______________________

Texture: tough, brittle, leathery, woody, soft, slimy, spongy, powdery, waxy, rubbery, watery, other: ______________________

______________________ Color ______________

Spore Color ______________________

Cap Shape & Features

bell-shaped conical convex cylindrical depressed

flat funnel-shaped umbilicate umbonate

Surface texture (warts, scales, slime, etc.) ______________________

____

Cap margin: smooth, inrolled, sinuous/wavy, other:

____

Color changes ______________________

Gill Features

◯ gills ◯ false gills ◯ pores ◯ teeth

Spacing

crowded close distant

Gill Attachment

adnexed
(narrowly attached)

adnate
(broadly attached)

decurrent
(running down the stalk)

free
(gills don't meet the stem)

sinuate
(notched)

sketch

Stem Shape

equal

club-shaped

bulbous

with cup
(volva)

tapering
downward

tapering
upward

sketch

Other Features

Stem attachment ◯ central ◯ offset ◯ lateral

Stem texture: fleshy, fibrous, hollow, solid, fibrillose, scaly, velvety, smooth, other: ______________________

Veil ◯ none ◯ partial ◯ universal

Veil texture: thin, cobwebby, fibrillose, slimy, other: ______________________

____

Annulus (ring)

◯ none ◯ collarlike

◯ skirtlike ◯ sheathlike

Volva

◯ none ◯ present

Other (bruising, bleeding, smell, etc.) ______________________

____

Mushroom Identification Log

Location

Site ______________________ **Date** ____________

◯ living tree ◯ leaf litter ◯ mulch ◯ other mushrooms
◯ dead tree or wood ◯ grass ◯ soil ◯ other ________

Type of tree(s) on or near ______________________

Forest type ______________________

Other (weather, lighting, humidity, etc.): ______________________

____

General

Size (cap, stem, & overall height): ______________________

Texture: tough, brittle, leathery, woody, soft, slimy, spongy, powdery, waxy, rubbery, watery, other: ____________

____________ Color ____________

Spore Color ______________________

Cap Shape & Features

bell-shaped, conical, convex, cylindrical, depressed

flat, funnel-shaped, umbilicate, umbonate

Surface texture (warts, scales, slime, etc.) ____________

____

Cap margin: smooth, inrolled, sinuous/wavy, other:

____

Color changes ______________________

Gill Features

◯ gills ◯ false gills ◯ pores ◯ teeth

Spacing

crowded close distant

Gill Attachment

adnexed
(narrowly attached)

adnate
(broadly attached)

decurrent
(running down the stalk)

free
(gills don't meet the stem)

sinuate
(notched)

sketch

Stem Shape

equal

club-shaped

bulbous

with cup
(volva)

tapering
downward

tapering
upward

sketch

Other Features

Stem attachment ◯ central ◯ offset ◯ lateral

Stem texture: fleshy, fibrous, hollow, solid, fibrillose, scaly, velvety, smooth, other: ____________

Veil ◯ none ◯ partial ◯ universal

Veil texture: thin, cobwebby, fibrillose, slimy, other: ____________

____

Annulus (ring)

◯ none ◯ collarlike

◯ skirtlike ◯ sheathlike

Volva

◯ none ◯ present

Other (bruising, bleeding, smell, etc.) ____________

____

Mushroom Identification Log

Location

Site ______________________ **Date** ________

◯ living tree ◯ leaf litter ◯ mulch ◯ other mushrooms
◯ dead tree or wood ◯ grass ◯ soil ◯ other ________

Type of tree(s) on or near ________

Forest type ________

Other (weather, lighting, humidity, etc.): ________

General

Size (cap, stem, & overall height): ________

Texture: tough, brittle, leathery, woody, soft, slimy, spongy, powdery, waxy, rubbery, watery, other: ________

________ Color ________

Spore Color ________

Cap Shape & Features

bell-shaped
conical
convex
cylindrical
depressed
flat
funnel-shaped
umbilicate
umbonate

Surface texture (warts, scales, slime, etc.) ________

Cap margin: smooth, inrolled, sinuous/wavy, other:

____

Color changes ________

Gill Features

◯ gills ◯ false gills ◯ pores ◯ teeth

Spacing

crowded
close
distant

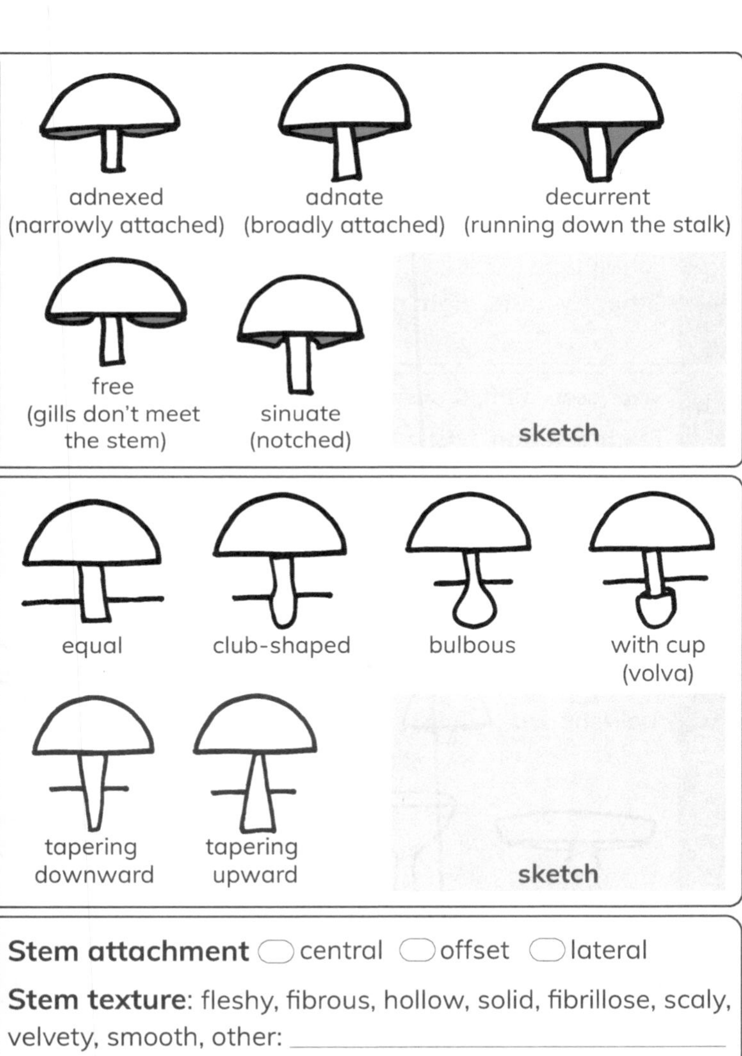

Other Features

Stem attachment ◯ central ◯ offset ◯ lateral

Stem texture: fleshy, fibrous, hollow, solid, fibrillose, scaly, velvety, smooth, other: ____________

Veil ◯ none ◯ partial ◯ universal

Veil texture: thin, cobwebby, fibrillose, slimy, other: ____________

____

Annulus (ring)

◯ none ◯ collarlike

◯ skirtlike ◯ sheathlike

Volva

◯ none ◯ present

Other (bruising, bleeding, smell, etc.) ____________

____

Mushroom Identification Log

Location

Site ______________________ **Date** ________

◯ living tree ◯ leaf litter ◯ mulch ◯ other mushrooms
◯ dead tree or wood ◯ grass ◯ soil ◯ other________

Type of tree(s) on or near________________________

Forest type______________________________

Other (weather, lighting, humidity, etc.): ____________

____

General

Size (cap, stem, & overall height): ______________

Texture: tough, brittle, leathery, woody, soft, slimy, spongy, powdery, waxy, rubbery, watery, other: ____________

__________________________ Color ________

Spore Color______________________________

Cap Shape & Features

bell-shaped
conical
convex
cylindrical
depressed
flat
funnel-shaped
umbilicate
umbonate

Surface texture (warts, scales, slime, etc.) ____________

____

Cap margin: smooth, inrolled, sinuous/wavy, other:

____

Color changes______________________________

Gill Features

◯ gills ◯ false gills ◯ pores ◯ teeth

Spacing

crowded
close
distant

Gill Attachment

adnexed
(narrowly attached)

adnate
(broadly attached)

decurrent
(running down the stalk)

free
(gills don't meet the stem)

sinuate
(notched)

sketch

Stem Shape

equal

club-shaped

bulbous

with cup
(volva)

tapering downward

tapering upward

sketch

Other Features

Stem attachment ◯ central ◯ offset ◯ lateral

Stem texture: fleshy, fibrous, hollow, solid, fibrillose, scaly, velvety, smooth, other: ____________________

Veil ◯ none ◯ partial ◯ universal

Veil texture: thin, cobwebby, fibrillose, slimy, other: __________

____

Annulus (ring)

◯ none ◯ collarlike

◯ skirtlike ◯ sheathlike

Volva

◯ none ◯ present

Other (bruising, bleeding, smell, etc.) ____________________

____

Mushroom Identification Log

Location

Site ______________________ **Date** __________

◯ living tree ◯ leaf litter ◯ mulch ◯ other mushrooms
◯ dead tree or wood ◯ grass ◯ soil ◯ other __________

Type of tree(s) on or near ______________________

Forest type ______________________

Other (weather, lighting, humidity, etc.): ______________________

General

Size (cap, stem, & overall height): ______________________

Texture: tough, brittle, leathery, woody, soft, slimy, spongy, powdery, waxy, rubbery, watery, other: ______________________

______________________ Color __________

Spore Color ______________________

Cap Shape & Features

bell-shaped

conical

convex

cylindrical

depressed

flat

funnel-shaped

umbilicate

umbonate

Surface texture (warts, scales, slime, etc.) ______________________

Cap margin: smooth, inrolled, sinuous/wavy, other:

____

Color changes ______________________

Gill Features

◯ gills ◯ false gills ◯ pores ◯ teeth

Spacing

crowded

close

distant

Gill Attachment

adnexed
(narrowly attached)

adnate
(broadly attached)

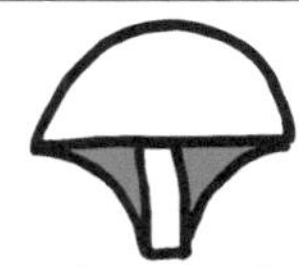

decurrent
(running down the stalk)

free
(gills don't meet the stem)

sinuate
(notched)

sketch

Stem Shape

equal

club-shaped

bulbous

with cup
(volva)

tapering downward

tapering upward

sketch

Other Features

Stem attachment ◯ central ◯ offset ◯ lateral

Stem texture: fleshy, fibrous, hollow, solid, fibrillose, scaly, velvety, smooth, other: ____________

Veil ◯ none ◯ partial ◯ universal

Veil texture: thin, cobwebby, fibrillose, slimy, other: ____________

____

Annulus (ring)

◯ none ◯ collarlike

◯ skirtlike ◯ sheathlike

Volva

◯ none ◯ present

Other (bruising, bleeding, smell, etc.) ____________

____

Mushroom Identification Log

Location

Site ______ **Date** ______

◯ living tree ◯ leaf litter ◯ mulch ◯ other mushrooms
◯ dead tree or wood ◯ grass ◯ soil ◯ other ______

Type of tree(s) on or near ______

Forest type ______

Other (weather, lighting, humidity, etc.): ______

General

Size (cap, stem, & overall height): ______

Texture: tough, brittle, leathery, woody, soft, slimy, spongy, powdery, waxy, rubbery, watery, other: ______

______ Color ______

Spore Color ______

Cap Shape & Features

bell-shaped

conical

convex

cylindrical

depressed

flat

funnel-shaped

umbilicate

umbonate

Surface texture (warts, scales, slime, etc.) ______

Cap margin: smooth, inrolled, sinuous/wavy, other:

______

Color changes ______

Gill Features

◯ gills ◯ false gills ◯ pores ◯ teeth

Spacing

crowded

close

distant

Gill Attachment

adnexed
(narrowly attached)

adnate
(broadly attached)

decurrent
(running down the stalk)

free
(gills don't meet the stem)

sinuate
(notched)

sketch

Stem Shape

equal

club-shaped

bulbous

with cup
(volva)

tapering downward

tapering upward

sketch

Other Features

Stem attachment ◯ central ◯ offset ◯ lateral

Stem texture: fleshy, fibrous, hollow, solid, fibrillose, scaly, velvety, smooth, other: ______________________

Veil ◯ none ◯ partial ◯ universal

Veil texture: thin, cobwebby, fibrillose, slimy, other: ______

____

Annulus (ring)

◯ none ◯ collarlike

◯ skirtlike ◯ sheathlike

Volva

◯ none ◯ present

Other (bruising, bleeding, smell, etc.) ______________________

____

Mushroom Identification Log

Location

Site ______________________ **Date** ____________

◯ living tree ◯ leaf litter ◯ mulch ◯ other mushrooms
◯ dead tree or wood ◯ grass ◯ soil ◯ other ____________

Type of tree(s) on or near ______________________

Forest type ______________________

Other (weather, lighting, humidity, etc.): ______________________

____

General

Size (cap, stem, & overall height): ______________________

Texture: tough, brittle, leathery, woody, soft, slimy, spongy, powdery, waxy, rubbery, watery, other: ______________________

______________________ Color ____________

Spore Color ______________________

Cap Shape & Features

bell-shaped

conical

convex

cylindrical

depressed

flat

funnel-shaped

umbilicate

umbonate

Surface texture (warts, scales, slime, etc.) ______________________

____

Cap margin: smooth, inrolled, sinuous/wavy, other:

____

Color changes ______________________

Gill Features

◯ gills ◯ false gills ◯ pores ◯ teeth

Spacing

crowded

close

distant

Gill Attachment

adnexed
(narrowly attached)

adnate
(broadly attached)

decurrent
(running down the stalk)

free
(gills don't meet the stem)

sinuate
(notched)

Stem Shape

equal

club-shaped

bulbous

with cup
(volva)

tapering
downward

tapering
upward

Other Features

Stem attachment ◯ central ◯ offset ◯ lateral

Stem texture: fleshy, fibrous, hollow, solid, fibrillose, scaly, velvety, smooth, other: ______________________

Veil ◯ none ◯ partial ◯ universal

Veil texture: thin, cobwebby, fibrillose, slimy, other: __________

____

Annulus (ring)

◯ none ◯ collarlike
◯ skirtlike ◯ sheathlike

Volva

◯ none ◯ present

Other (bruising, bleeding, smell, etc.) ______________________

____

Mushroom Identification Log

Location

Site ______________________ **Date** __________

◯ living tree ◯ leaf litter ◯ mulch ◯ other mushrooms
◯ dead tree or wood ◯ grass ◯ soil ◯ other__________

Type of tree(s) on or near______________________

Forest type______________________

Other (weather, lighting, humidity, etc.): ______________________

____

General

Size (cap, stem, & overall height): ______________________

Texture: tough, brittle, leathery, woody, soft, slimy, spongy, powdery, waxy, rubbery, watery, other: __________

______________________ Color __________

Spore Color______________________

Cap Shape & Features

bell-shaped

conical

convex

cylindrical

depressed

flat

funnel-shaped

umbilicate

umbonate

Surface texture (warts, scales, slime, etc.) __________

____

Cap margin: smooth, inrolled, sinuous/wavy, other:

____

Color changes______________________

Gill Features

◯ gills ◯ false gills ◯ pores ◯ teeth

Spacing

crowded

close

distant

Gill Attachment

adnexed
(narrowly attached)

adnate
(broadly attached)

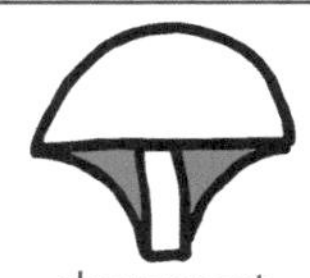
decurrent
(running down the stalk)

free
(gills don't meet the stem)

sinuate
(notched)

sketch

Stem Shape

equal

club-shaped

bulbous

with cup
(volva)

tapering
downward

tapering
upward

sketch

Other Features

Stem attachment ⬭ central ⬭ offset ⬭ lateral

Stem texture: fleshy, fibrous, hollow, solid, fibrillose, scaly, velvety, smooth, other: ______________________

Veil ⬭ none ⬭ partial ⬭ universal

Veil texture: thin, cobwebby, fibrillose, slimy, other: ______
____

Annulus (ring)
⬭ none ⬭ collarlike
⬭ skirtlike ⬭ sheathlike

Volva
⬭ none ⬭ present

Other (bruising, bleeding, smell, etc.) ______________________
____

Mushroom Identification Log

Location

Site ____________________ **Date** ____________

◯ living tree ◯ leaf litter ◯ mulch ◯ other mushrooms
◯ dead tree or wood ◯ grass ◯ soil ◯ other ____________

Type of tree(s) on or near ____________________

Forest type ____________________

Other (weather, lighting, humidity, etc.): ____________________

General

Size (cap, stem, & overall height): ____________________

Texture: tough, brittle, leathery, woody, soft, slimy, spongy, powdery, waxy, rubbery, watery, other: ____________________

____________________ Color ____________

Spore Color ____________________

Cap Shape & Features

bell-shaped

conical

convex

cylindrical

depressed

flat

funnel-shaped

umbilicate

umbonate

Surface texture (warts, scales, slime, etc.) ____________________

Cap margin: smooth, inrolled, sinuous/wavy, other:

____

Color changes ____________________

Gill Features

◯ gills ◯ false gills ◯ pores ◯ teeth

Spacing

crowded

close

distant

Gill Attachment

adnexed
(narrowly attached)

adnate
(broadly attached)

decurrent
(running down the stalk)

free
(gills don't meet the stem)

sinuate
(notched)

sketch

Stem Shape

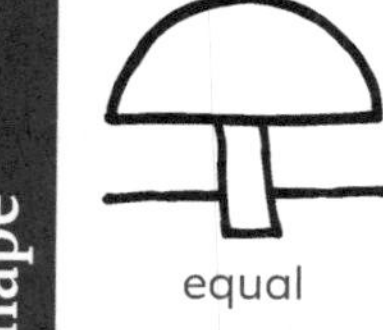
equal

club-shaped

bulbous

with cup
(volva)

tapering
downward

tapering
upward

sketch

Other Features

Stem attachment ◯ central ◯ offset ◯ lateral

Stem texture: fleshy, fibrous, hollow, solid, fibrillose, scaly, velvety, smooth, other: ____________________

Veil ◯ none ◯ partial ◯ universal

Veil texture: thin, cobwebby, fibrillose, slimy, other: ________

____

Annulus (ring)

◯ none ◯ collarlike
◯ skirtlike ◯ sheathlike

Volva

◯ none ◯ present

Other (bruising, bleeding, smell, etc.) ____________________

____

Mushroom Identification Log

Location

Site ______________________ **Date** ____________

◯ living tree ◯ leaf litter ◯ mulch ◯ other mushrooms
◯ dead tree or wood ◯ grass ◯ soil ◯ other ____________

Type of tree(s) on or near ______________________

Forest type ______________________

Other (weather, lighting, humidity, etc.): ______________________

____

General

Size (cap, stem, & overall height): ______________________

Texture: tough, brittle, leathery, woody, soft, slimy, spongy, powdery, waxy, rubbery, watery, other: ____________

______________________ Color ____________

Spore Color ______________________

Cap Shape & Features

bell-shaped, conical, convex, cylindrical, depressed

flat, funnel-shaped, umbilicate, umbonate

Surface texture (warts, scales, slime, etc.) ____________

____

Cap margin: smooth, inrolled, sinuous/wavy, other:

____

Color changes ______________________

Gill Features

◯ gills ◯ false gills ◯ pores ◯ teeth

Spacing

crowded, close, distant

Gill Attachment

adnexed (narrowly attached)

adnate (broadly attached)

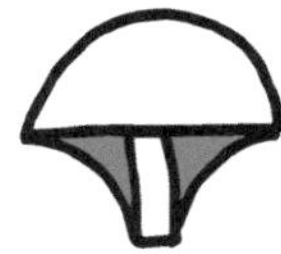
decurrent (running down the stalk)

free (gills don't meet the stem)

sinuate (notched)

sketch

Stem Shape

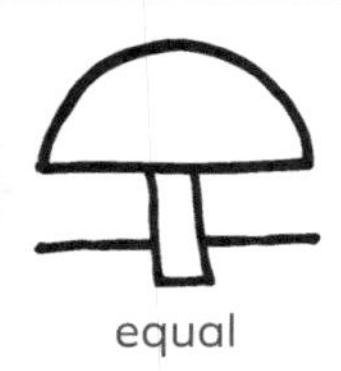
equal

club-shaped

bulbous

with cup (volva)

tapering downward

tapering upward

sketch

Other Features

Stem attachment ◯ central ◯ offset ◯ lateral

Stem texture: fleshy, fibrous, hollow, solid, fibrillose, scaly, velvety, smooth, other: ______

Veil ◯ none ◯ partial ◯ universal

Veil texture: thin, cobwebby, fibrillose, slimy, other: ______

______

Annulus (ring)

◯ none ◯ collarlike

◯ skirtlike ◯ sheathlike

Volva

◯ none ◯ present

Other (bruising, bleeding, smell, etc.) ______

______

Mushroom Identification Log

Location

Site ________ **Date** ________

◯ living tree ◯ leaf litter ◯ mulch ◯ other mushrooms
◯ dead tree or wood ◯ grass ◯ soil ◯ other________

Type of tree(s) on or near________

Forest type________

Other (weather, lighting, humidity, etc.): ________

General

Size (cap, stem, & overall height): ________

Texture: tough, brittle, leathery, woody, soft, slimy, spongy, powdery, waxy, rubbery, watery, other: ________

Color ________

Spore Color________

Cap Shape & Features

bell-shaped, conical, convex, cylindrical, depressed

flat, funnel-shaped, umbilicate, umbonate

Surface texture (warts, scales, slime, etc.) ________

Cap margin: smooth, inrolled, sinuous/wavy, other:

Color changes________

Gill Features

◯ gills ◯ false gills ◯ pores ◯ teeth

Spacing

crowded, close, distant

Gill Attachment

adnexed
(narrowly attached)

adnate
(broadly attached)

decurrent
(running down the stalk)

free
(gills don't meet the stem)

sinuate
(notched)

sketch

Stem Shape

equal

club-shaped

bulbous

with cup
(volva)

tapering downward

tapering upward

sketch

Other Features

Stem attachment ◯ central ◯ offset ◯ lateral

Stem texture: fleshy, fibrous, hollow, solid, fibrillose, scaly, velvety, smooth, other: ____________________

Veil ◯ none ◯ partial ◯ universal

Veil texture: thin, cobwebby, fibrillose, slimy, other: ____________________

____

Annulus (ring)

◯ none ◯ collarlike

◯ skirtlike ◯ sheathlike

Volva

◯ none ◯ present

Other (bruising, bleeding, smell, etc.) ____________________

____

Mushroom Identification Log

Location

Site ______________________ **Date** ____________

◯ living tree ◯ leaf litter ◯ mulch ◯ other mushrooms
◯ dead tree or wood ◯ grass ◯ soil ◯ other ________

Type of tree(s) on or near ______________________

Forest type ______________________

Other (weather, lighting, humidity, etc.): ______________________

____

General

Size (cap, stem, & overall height): ______________________

Texture: tough, brittle, leathery, woody, soft, slimy, spongy, powdery, waxy, rubbery, watery, other: ______________

______________________ Color ____________

Spore Color ______________________

Cap Shape & Features

bell-shaped

conical

convex

cylindrical

depressed

flat

funnel-shaped

umbilicate

umbonate

Surface texture (warts, scales, slime, etc.) ______________

____

Cap margin: smooth, inrolled, sinuous/wavy, other:

____

Color changes ______________________

Gill Features

◯ gills ◯ false gills ◯ pores ◯ teeth

Spacing

crowded

close

distant

Gill Attachment

adnexed
(narrowly attached)

adnate
(broadly attached)

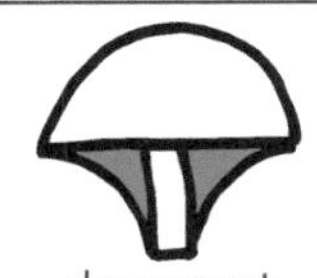

decurrent
(running down the stalk)

free
(gills don't meet the stem)

sinuate
(notched)

sketch

Stem Shape

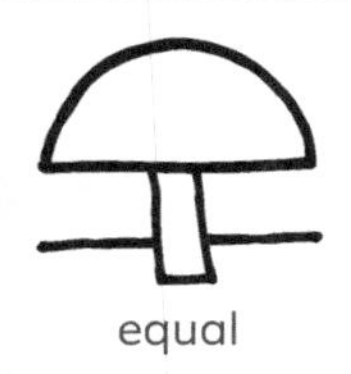

equal

club-shaped

bulbous

with cup
(volva)

tapering
downward

tapering
upward

sketch

Other Features

Stem attachment ◯ central ◯ offset ◯ lateral

Stem texture: fleshy, fibrous, hollow, solid, fibrillose, scaly, velvety, smooth, other: __________

Veil ◯ none ◯ partial ◯ universal

Veil texture: thin, cobwebby, fibrillose, slimy, other: __________

____

Annulus (ring)

◯ none ◯ collarlike

◯ skirtlike ◯ sheathlike

Volva

◯ none ◯ present

Other (bruising, bleeding, smell, etc.) __________

____

Mushroom Identification Log

Location

Site ______________________ **Date** ____________

◯ living tree ◯ leaf litter ◯ mulch ◯ other mushrooms
◯ dead tree or wood ◯ grass ◯ soil ◯ other ____________

Type of tree(s) on or near ______________________

Forest type ______________________

Other (weather, lighting, humidity, etc.): ______________________

____

General

Size (cap, stem, & overall height): ______________________

Texture: tough, brittle, leathery, woody, soft, slimy, spongy, powdery, waxy, rubbery, watery, other: ______________________

______________________ Color ____________

Spore Color ______________________

Cap Shape & Features

bell-shaped

conical

convex

cylindrical

depressed

flat

funnel-shaped

umbilicate

umbonate

Surface texture (warts, scales, slime, etc.) ______________________

____

Cap margin: smooth, inrolled, sinuous/wavy, other:

____

Color changes ______________________

Gill Features

◯ gills ◯ false gills ◯ pores ◯ teeth

Spacing

crowded

close

distant

Gill Attachment

adnexed
(narrowly attached)

adnate
(broadly attached)

decurrent
(running down the stalk)

free
(gills don't meet the stem)

sinuate
(notched)

sketch

Stem Shape

equal

club-shaped

bulbous

with cup
(volva)

tapering
downward

tapering
upward

sketch

Other Features

Stem attachment ◯ central ◯ offset ◯ lateral

Stem texture: fleshy, fibrous, hollow, solid, fibrillose, scaly, velvety, smooth, other: ______________________

Veil ◯ none ◯ partial ◯ universal

Veil texture: thin, cobwebby, fibrillose, slimy, other: ________

____

Annulus (ring)

◯ none ◯ collarlike
◯ skirtlike ◯ sheathlike

Volva

◯ none ◯ present

Other (bruising, bleeding, smell, etc.) ______________

____

Mushroom Identification Log

Location

Site ______________________ **Date** __________

◯ living tree ◯ leaf litter ◯ mulch ◯ other mushrooms
◯ dead tree or wood ◯ grass ◯ soil ◯ other __________

Type of tree(s) on or near ______________________

Forest type ______________________

Other (weather, lighting, humidity, etc.): ______________________

____

General

Size (cap, stem, & overall height): ______________________

Texture: tough, brittle, leathery, woody, soft, slimy, spongy, powdery, waxy, rubbery, watery, other: __________

__________ Color __________

Spore Color ______________________

Cap Shape & Features

bell-shaped
conical
convex
cylindrical
depressed
flat
funnel-shaped
umbilicate
umbonate

Surface texture (warts, scales, slime, etc.) __________

____

Cap margin: smooth, inrolled, sinuous/wavy, other:

____

Color changes ______________________

Gill Features

◯ gills ◯ false gills ◯ pores ◯ teeth

Spacing

crowded
close
distant

Gill Attachment

adnexed
(narrowly attached)

adnate
(broadly attached)

decurrent
(running down the stalk)

free
(gills don't meet the stem)

sinuate
(notched)

sketch

Stem Shape

equal

club-shaped

bulbous

with cup
(volva)

tapering
downward

tapering
upward

sketch

Other Features

Stem attachment ◯ central ◯ offset ◯ lateral

Stem texture: fleshy, fibrous, hollow, solid, fibrillose, scaly, velvety, smooth, other: ______

Veil ◯ none ◯ partial ◯ universal

Veil texture: thin, cobwebby, fibrillose, slimy, other: ______

______

Annulus (ring)

◯ none ◯ collarlike
◯ skirtlike ◯ sheathlike

Volva

◯ none ◯ present

Other (bruising, bleeding, smell, etc.) ______

______

Mushroom Identification Log

Location

Site ______ **Date** ______

◯ living tree ◯ leaf litter ◯ mulch ◯ other mushrooms
◯ dead tree or wood ◯ grass ◯ soil ◯ other ______

Type of tree(s) on or near ______

Forest type ______

Other (weather, lighting, humidity, etc.): ______

General

Size (cap, stem, & overall height): ______

Texture: tough, brittle, leathery, woody, soft, slimy, spongy, powdery, waxy, rubbery, watery, other: ______

______ Color ______

Spore Color ______

Cap Shape & Features

bell-shaped

conical

convex

cylindrical

depressed

flat

funnel-shaped

umbilicate

umbonate

Surface texture (warts, scales, slime, etc.) ______

Cap margin: smooth, inrolled, sinuous/wavy, other:

______

Color changes ______

Gill Features

◯ gills ◯ false gills ◯ pores ◯ teeth

Spacing

crowded

close

distant

Gill Attachment

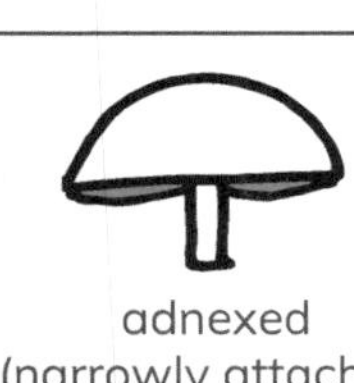

adnexed
(narrowly attached)

adnate
(broadly attached)

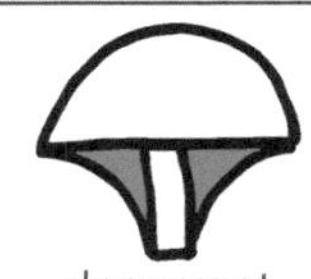

decurrent
(running down the stalk)

free
(gills don't meet the stem)

sinuate
(notched)

sketch

Stem Shape

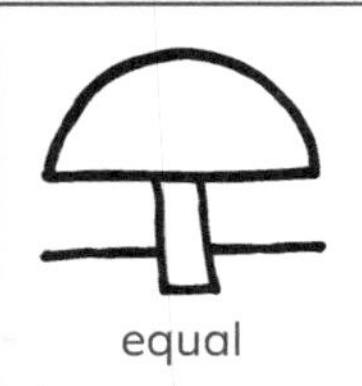

equal

club-shaped

bulbous

with cup
(volva)

tapering
downward

tapering
upward

sketch

Other Features

Stem attachment ◯ central ◯ offset ◯ lateral

Stem texture: fleshy, fibrous, hollow, solid, fibrillose, scaly, velvety, smooth, other: ____________________

Veil ◯ none ◯ partial ◯ universal

Veil texture: thin, cobwebby, fibrillose, slimy, other: __________

____

Annulus (ring)
◯ none ◯ collarlike
◯ skirtlike ◯ sheathlike

Volva
◯ none ◯ present

Other (bruising, bleeding, smell, etc.) ____________________

____

Mushroom Identification Log

Location

Site ____________________ **Date** __________

◯ living tree ◯ leaf litter ◯ mulch ◯ other mushrooms
◯ dead tree or wood ◯ grass ◯ soil ◯ other __________

Type of tree(s) on or near ____________________

Forest type ____________________

Other (weather, lighting, humidity, etc.): ____________________

____

General

Size (cap, stem, & overall height): ____________________

Texture: tough, brittle, leathery, woody, soft, slimy, spongy, powdery, waxy, rubbery, watery, other: __________

____________________ Color __________

Spore Color ____________________

Cap Shape & Features

bell-shaped
conical
convex
cylindrical
depressed
flat
funnel-shaped
umbilicate
umbonate

Surface texture (warts, scales, slime, etc.) ____________________

____

Cap margin: smooth, inrolled, sinuous/wavy, other:

____

Color changes ____________________

Gill Features

◯ gills ◯ false gills ◯ pores ◯ teeth

Spacing

crowded
close
distant

Gill Attachment

adnexed
(narrowly attached)

adnate
(broadly attached)

decurrent
(running down the stalk)

free
(gills don't meet the stem)

sinuate
(notched)

sketch

Stem Shape

equal

club-shaped

bulbous

with cup
(volva)

tapering
downward

tapering
upward

sketch

Other Features

Stem attachment ◯ central ◯ offset ◯ lateral

Stem texture: fleshy, fibrous, hollow, solid, fibrillose, scaly, velvety, smooth, other: ______________________

Veil ◯ none ◯ partial ◯ universal

Veil texture: thin, cobwebby, fibrillose, slimy, other: ________

____

Annulus (ring)

◯ none ◯ collarlike

◯ skirtlike ◯ sheathlike

Volva

◯ none ◯ present

Other (bruising, bleeding, smell, etc.) ______________

____

Mushroom Identification Log

Location

Site ______________________ **Date** ______________

◯ living tree ◯ leaf litter ◯ mulch ◯ other mushrooms

◯ dead tree or wood ◯ grass ◯ soil ◯ other ______

Type of tree(s) on or near ______________________

Forest type ______________________

Other (weather, lighting, humidity, etc.): ______________________

____

General

Size (cap, stem, & overall height): ______________________

Texture: tough, brittle, leathery, woody, soft, slimy, spongy, powdery, waxy, rubbery, watery, other: ______________

______________________ Color ______________

Spore Color ______________________

Cap Shape & Features

bell-shaped

conical

convex

cylindrical

depressed

flat

funnel-shaped

umbilicate

umbonate

Surface texture (warts, scales, slime, etc.) ______________

____

Cap margin: smooth, inrolled, sinuous/wavy, other:

____

Color changes ______________________

Gill Features

◯ gills ◯ false gills ◯ pores ◯ teeth

Spacing

crowded

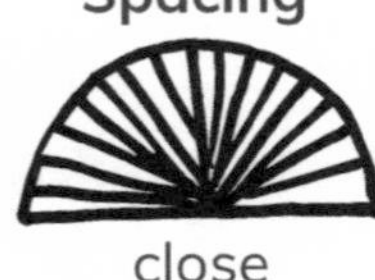

close

distant

Gill Attachment

adnexed (narrowly attached) | adnate (broadly attached) | decurrent (running down the stalk)

free (gills don't meet the stem) | sinuate (notched)

sketch

Stem Shape

equal | club-shaped | bulbous | with cup (volva)

tapering downward | tapering upward

sketch

Other Features

Stem attachment ◯ central ◯ offset ◯ lateral

Stem texture: fleshy, fibrous, hollow, solid, fibrillose, scaly, velvety, smooth, other: ____________

Veil ◯ none ◯ partial ◯ universal

Veil texture: thin, cobwebby, fibrillose, slimy, other: ____________

____

Annulus (ring)

◯ none ◯ collarlike
◯ skirtlike ◯ sheathlike

Volva

◯ none ◯ present

Other (bruising, bleeding, smell, etc.) ____________

____

Mushroom Identification Log

Location

Site ______________________ **Date** __________

◯ living tree ◯ leaf litter ◯ mulch ◯ other mushrooms

◯ dead tree or wood ◯ grass ◯ soil ◯ other__________

Type of tree(s) on or near__________________________

Forest type______________________________________

Other (weather, lighting, humidity, etc.): ______________

____

General

Size (cap, stem, & overall height): ___________________

Texture: tough, brittle, leathery, woody, soft, slimy, spongy, powdery, waxy, rubbery, watery, other: ______________

______________________________ Color ______________

Spore Color_______________________________________

Cap Shape & Features

bell-shaped

conical

convex

cylindrical

depressed

flat

funnel-shaped

umbilicate

umbonate

Surface texture (warts, scales, slime, etc.) ____________

____

Cap margin: smooth, inrolled, sinuous/wavy, other:

____

Color changes_____________________________________

Gill Features

◯ gills ◯ false gills ◯ pores ◯ teeth

Spacing

crowded

close

distant

Gill Attachment

adnexed
(narrowly attached)

adnate
(broadly attached)

decurrent
(running down the stalk)

free
(gills don't meet the stem)

sinuate
(notched)

sketch

Stem Shape

equal

club-shaped

bulbous

with cup
(volva)

tapering
downward

tapering
upward

sketch

Other Features

Stem attachment ◯ central ◯ offset ◯ lateral

Stem texture: fleshy, fibrous, hollow, solid, fibrillose, scaly, velvety, smooth, other: ____________________

Veil ◯ none ◯ partial ◯ universal

Veil texture: thin, cobwebby, fibrillose, slimy, other: ________

____

Annulus (ring)

◯ none ◯ collarlike

◯ skirtlike ◯ sheathlike

Volva

◯ none ◯ present

Other (bruising, bleeding, smell, etc.) ____________

____

Observations

Observations

Observations

Observations

Observations

Observations

Observations

Made in United States
Troutdale, OR
09/10/2024